Marketing of engineering services

ENGINEERING MANAGEMENT

Other titles in the series

Civil engineering insurance and bonding, P. Madge

Construction planning, R. H. Neale and D. E. Neale

Engineering contracts, S. H. Wearne

Management of design offices, P. A. Rutter (Ed)

Managing people, A. S. Martin and F. Grover

Project evaluation, R. K. Corrie (Ed)

Financial control, N. M. L. Barnes (Ed)

Control of engineering projects, S. H. Wearne et al

ENGINEERING MANAGEMENT

Marketing of engineering services

Brian Scanlon, BSc, MSc, MIMC

Thomas Telford, London

Published by Thomas Telford Ltd, Thomas Telford House,
1 Heron Quay, London E14 9XF

First published 1988

British Library Cataloguing in Publication Data
Scanlon, Brian
Marketing of engineering services.
1. Engineering. Marketing
I. Title II. Series
620'.0068'8

ISBN: 0 7277 1348 5

Typeset in Great Britain by MHL Typesetting Limited, Coventry
Printed and bound in Great Britain by Billings & Sons, Limited, Worcester

Contents

1 **Introduction** 1
Perspective of marketing within civil engineering; what is marketing?; why is it necessary?; structure of this book

2 **The structured approach** 5
The need for structure; the key structural elements in marketing

3 **Client orientation** 7
Basic postures; enquiry-driven stance; sales-driven stance; market-driven stance; customer engineering; evolution of business controls; patterns of customer profitability; engineering profitability

4 **Strategic planning** 13
Links with marketing; scope of strategic planning; hierarchy of business activities; approach to planning; organizational issues; coordination; investment planning

5 **Operational planning** 21
Starting a marketing initiative; budgeting; fast food illustration; consumer goods illustration; implications for civil engineering activities; characteristics of marketing in civil engineering; establishing a marketing budget; operational management; links with selling budgets

6 **Management of intelligence** 29
Basic requirements; categorization; scope; control; acquisition; dissemination; interactive needs; summary of priorities

7 **Promotional activities** 34
Introduction; image; house style; public relations; the role of the media; publications; advertising; summary of priorities

8 **Presentations** 44
Objectives; legal profession experience; importance of preparation; critical role of conclusions; general format for presentations; managing the props; managing the audience; performance aspects; summary and golden rules

9 **Submissions** 55
Objectives; the use of proposals; preparation as part of the research process; bid strategies; sequence of preparation; some general formats; speculation proposals; submission document team; summary and golden rules

10 **People** 69
People orientation; essential prerequisites for success; attitude to marketing; performance measurement; training

11 **Segmentation** 77
Putting marketing into practice; concept of segmentation; dimensions and scope of segmentation; segmentation as an analytic tool

12 **Performance analysis** 83
Measurement of performance; corporate measures of performance; external and internal perspectives; setting corporate performance targets; use of ROTA curves; strategic development; portfolio analysis

13 **The consulting engineer** 94
Role; evolution of the profession; future trends; the key results areas; key marketing themes; essential staff work

14 **The contractor** 100
Role; evolution of the business; the critical impact of the public sector; future trends; the key results areas; essential staff work; the prospects for venturing

15 **The public service authority** 106
Role; evolution of the profession; future trends; the key results areas; key marketing themes; essential staff work

16 **Conclusions** 110
Contrasting characteristics; pattern for career development

1 Introduction

Perspective of marketing within civil engineering

Marketing as a management activity is still a comparatively recent phenomenon in the United Kingdom. During the 1960s and the early 1970s it tended to be regarded as a technique rather than an activity and this helped to create an aura of mystique around the subject. It is, however, a fundamental activity within business and when properly addressed is amenable to the sort of logical and structured approach that engineers use so readily in other business areas.

Civil engineers, in general, seem to have a sceptical view about the merits of marketing and the role they themselves should play. A recent survey in a British construction company about attitudes to marketing produced surprising results. Typically, it was seen as:

- an activity carried out by 'shallow' people
- an activity which does not represent 'real' work
- something which can never provide job satisfaction for a 'normal' engineer
- an activity best left to other people
- a role typically filled by executives moving 'sideways'
- something foreign competitors tend to do well because they find it easier to commit the funds to do it.

Despite this scepticism, many engineers do recognize good marketing and attribute value to it when they see it done well by their competitors. Paradoxically, recognizing marketing skills in others appears to do little in helping many engineers to conceive how it should be applied to their own organizations, or to justify appropriate resources for it.

What is marketing?

The problems with marketing start with definitions. Stated simply it is the concept of matching services to wants in the marketplace. This concept is much more than merely selling. It is the agent for promoting the sort of ongoing healthy change that enables an organization to modify its activities in line with shifts in its marketplace and business environment.

Rather than use a tight technical definition, it is more valuable to look at marketing as an exercise in orchestration.

- As shown in Fig. 1, all businesses can be regarded as having three common components:

  - *Markets:* which provide opportunities to supply products or services and so generate revenues.
  - *Activities:* which are constrained by resources and limit business transactions to specific products and services.
  - *Competition:* which sets price and service expectations and so tends to limit further the sales opportunities.

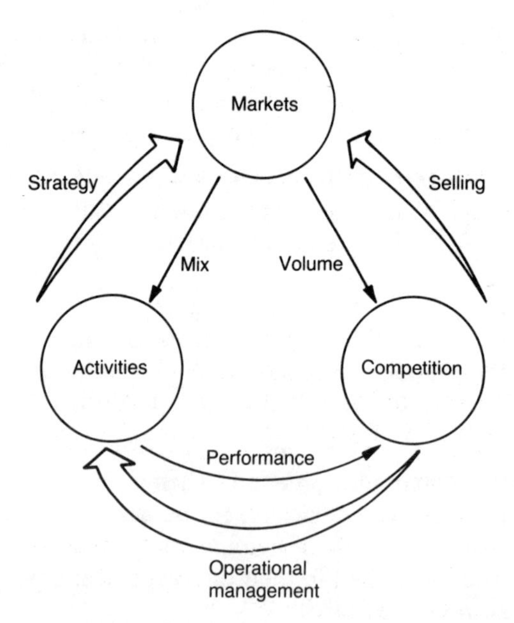

Fig. 1. Simplified model of key business relationships

- In looking at the relationships between these three main components of business it can be seen that:
  - Strategy formulation is the process of relating markets to activities. Its output is to determine something called *mix*. That is the specific profile of services/products and clients/customers that make up the marketplace for the organization.
  - Selling is the process of creating *volume* by placing services/products into the marketplace in a competitive environment.
  - Operational management is charged with producing adequate services/products in a competitive environment and its success in doing this is a measure of *performance*.
- *Marketing* is the process of orchestrating these dynamic relationships. Its main purposes are:
  - to develop services or products that will meet client or customer needs in a profitable and competitive way;
  - to communicate an awareness amongst clients or customers in the marketplace of the services or products that are offered.

Why is it necessary?

Without marketing new services or products do not get developed to meet the changing needs of a marketplace. Without marketing prospective clients or customers are unaware of services or products that they might need in developing their own business activities or in meeting their own performance objectives.

Structure of the book

This book is essentially in four parts:

- *Framework for marketing* (chapters 2–5)
  This puts forward a structured approach to marketing which is essential if professional standards are to be developed.

- *Practical approaches* (chapters 6–10)
  This tackles some of the practical management and presentational skills that have a special relevance in marketing. It attempts to lay the groundwork for further development. Proficiency in these aspects of marketing, however, only comes with practice.

- *Special techniques* (chapters 11 and 12)
  This addresses some of the techniques of analysis that are vital in long term planning and strategic evaluations. These are essential components in producing successful marketing plans.
- *Professional aspects* (chapters 13–15)
  This draws out the special circumstances that will confront engineers working in the different environments of consultancy practices, contractors, and public service authorities.

2 The structured approach

The need for structure

Marketing, when viewed from the outside, can appear to be a very confusing activity. Immediate questions arise regarding what tasks ought to be undertaken and, probably more importantly, how one establishes whether a task has been done or done well. Part of the confusion arises because marketing touches so many facets of business operations that managers can often feel either inadequate or that they are encroaching in part on other areas of responsibility.

- It is important to realize at the outset that:

  - marketing is a general responsibility for all management, even when specific marketing resources appear to be dedicated to the task;
  - marketing is not a form of personality cult to which only a few types of people can aspire.

Marketing, as with all other forms of management activity, does have an intrinsic structure and, furthermore, is very amenable to a structured approach.

The key structural elements in marketing

- One way of structuring an approach to marketing is to recognize that:

  - the whole purpose is to understand the needs of the market place, i.e. *client orientation*
  - plans need to be developed to produce appropriate services or products for the marketplace and its anticipated growth or evolution, i.e. *strategic planning*
  - plans are needed to provide specific services or products to

prospective clients or customers in profitable, competitive and compellingly attractive ways, i.e. *operational planning*

- o every transaction results from learning about specific needs and being able to provide an appropriate response, i.e. the *management of intelligence*
- o it is more effective if it can be arranged for prospective clients or customers to enquire about how their needs can be serviced by corresponding with leading organizations in the field, i.e. the world of *promotional activities*.

- The key structural elements of marketing put forward in this monograph are simply:

- o client orientations
- o strategic planning
- o operational planning
- o management of intelligence
- o promotional activities.

Each of these points is treated separately in later chapters.

Whilst a structured approach is essential, in that it provides a logical framework of approach, it should be noted that there is no accepted fixed approach to marketing. It is not a precise scientific tool. It needs to be flexible to suit the ever changing needs of the marketplace and most importantly has to encourage the practitioner to make the critical creative leap in providing the marketplace with new and exciting developments.

3 Client orientation

Basic postures

Marketing, which is primarily concerned with providing society with its needs for products and services in a cost-effective and profitable way, must start with the customer or the client.

- There are three postures an organization can adopt with regard to its customers or clients. These are:

  - an enquiry-driven stance
  - a sales-driven stance
  - a marketing-driven stance.

Each of these is explored in more detail.

Enquiry-driven stance

If an organization waits until prospective customers or clients identify it as a possible supplier and only responds to the enquiries that result, it has adopted an enquiry-driven stance.

Every organization gets some business in this way. This is very often one of the benefits of being a leader in the field, and leadership should be sought specifically to engineer just these sort of enquiries as they put organizations into a strong competitive position very early and for very little cost.

Few organizations get sufficient business to survive this way, and those who do are likely to be threatened in the near future as business relationships and the needs of society change. Reliance on this form of business development is best described as non-marketing.

- This stance is often the trap that ensnares the following groups:

- o professional practices, who feel that their professional code of ethics preclude any other marketing posture;
- o public servants, particularly those involved in the management of utilities, who feel that other forms of marketing are inconsistent with public sector responsibilities and accountabilities;
- o market leaders, who have established that most business opportunities come to them, and that few seem to be generated by other initiatives.

The biggest failing of this stance is that it reduces customer contact to a minimum. This develops into a detachment from the marketplace, and is not compatible with policies to maximize sales and earnings opportunities.

Sales-driven stance

When an organization attempts to sell whatever it happens to be capable of producing, it is adopting a sales-driven stance.

Many organizations see the ultimate ability in marketing as the ability to do just this task. It certainly focuses attention on the immediate situation and will go some way to improve performance. However, its fundamental weakness is that it neglects the question of what customers and clients really want, and what they will expect in the future.

This stance leads to vulnerability in the long term as fundamental changes in market requirements take place.

- Organizations that are most susceptible to this trap are usually:

- o companies with high levels of operational overheads, where there is constant pressure to keep resources fully utilized and where there are major investment barriers to developing new products or services;
- o organizations with large field sales activities which are overly focused on short term revenue generation through incentives or bonus systems.

Market-driven stance

When an organization attempts to identify what the marketplace needs and then provide it credibly and profitably it is adopting a market-driven stance.

Whilst this stance is common for organizations supplying fast

	Stance		
	Enquiry-driven	Sales-driven	Market-driven
Guiding principle:	'We're ready when the customer wants us'	'We sell what the company makes'	'We supply what the market needs'
Business emphasis:	Universal awareness	Cost containment	Growth and innovation
Vulnerability:	Medium term – to missed opportunities and shifting fashions	Long term – to changing needs	Short term – to product development success

Fig. 2. Client orientation

moving consumer goods, there are few organizations outside that sector that adopt this approach or even feel comfortable to consider it. One of the biggest barriers is simply cost. This stance is more expensive to adopt than the others and it is more difficult to manage effectively. However, no matter how daunting or unattractive it might appear to be to embark on a programme to create a market-driven stance, it is the only posture capable of providing the basis for long term business growth and development.

However, simply adopting this stance is not sufficient to ensure future business success. There are many organizations with this approach that are not commercially successful. When this is the case, it is not the concept that is wrong or expensive. It is the effectiveness of the marketing itself that is at fault.

The key issues regarding the three basic postures to marketing are summarized in Fig. 2.

Customer engineering

It often comes as a surprise when it is realized that customer or client behaviour can be engineered to produce a desired result. This process has nothing to do with any unacceptable business practices, such as subliminal suggestion or gifts or entertainment, but is based on a quantified or statistical understanding of the groups that provide an organization with revenues.

However, the concept of customer engineering needs some introduction.

Evolution of business controls

The first business organizations merely set out to establish whether an overall profit or loss was made on operations. In the course of time it was felt necessary to exercise a greater degree of control over operations so that a more predictable outcome would result. These moves led to the first type of business control which was process costing. These controls were felt to be adequate probably up to the time of the First World War.

However, organizations do not sell 'standard-hours' of activity; they sell products or services. With the growth between the Wars of the consumer durables businesses a new impetus was given to improve cost controls. This led to the introduction of product costing techniques which now dominate modern management practice worldwide.

There is now one further stage of development that needs to be tackled. Neither processes nor products or services produce a revenue directly. Revenues are generated by customers or clients. It therefore makes more sense to treat the customer as the ultimate profit centre and attribute all costs to servicing their individual needs. The cost allocation principles involved are no different from those involved in developing product or service costs, and the procedure is equally valid even if initially it does look very unwieldy.

Patterns of customer profitability

When the concepts of customer cost accounting were first applied to produce analyses of contributions to profits by customers, there were some very surprising results. Many organizations discovered that their profile of account profitabilities were distributed randomly with a slight skew towards an overall profit. A typical profile is given schematically in Fig. 3.

Most organizations will experience this phenomenon and most will fail totally to understand its significance. If customer account profitabilities exhibit a skew-normal distribution it means that the customers themselves control the level of profits, not the actions of the organization. This conclusion is fundamental but not obvious and needs to be absorbed before its full impact can be appreciated. To make the point another way, unless organizations set out to constrain customer or client behaviour into predetermined patterns, the net impact of trading on a business is that overall results are

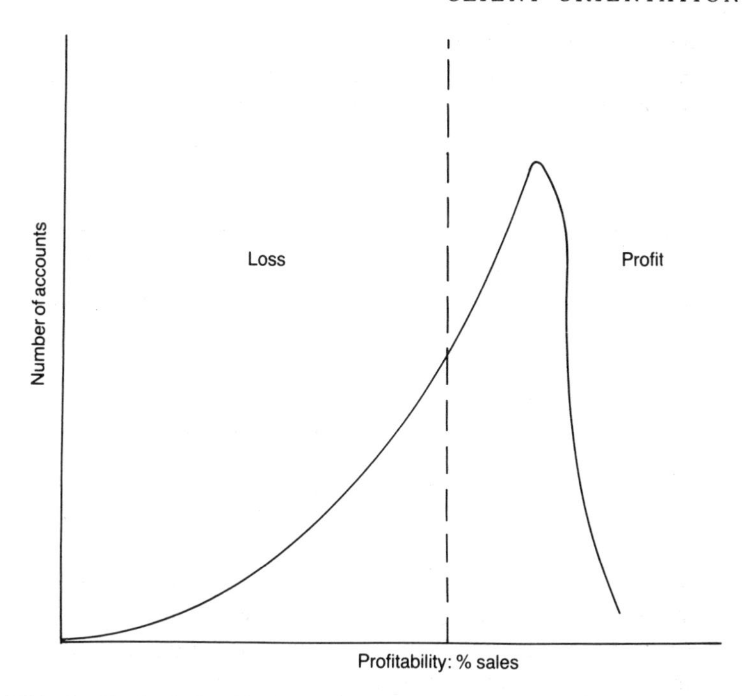

Fig. 3. Typical distribution of customer account profitabilities

simply a statistical average of the range of trading relationships open to the market at large.

This effect is dramatic in businesses with thousands of accounts, each of which may be active on a weekly basis. In civil engineering there will be very few accounts and the trading frequencies will be low, but the effects are still there and can be observed when groups of like situations are lumped together.

Engineering profitability

The concept of customer engineering comes alive in civil engineering when bidding strategies are reviewed. Client orientation should not be confined to understanding the needs of clients; it should also include an appreciation of how valuable clients can be and how they should be managed.

When an organization decides to pursue a project it assumes, implicitly, that the costs of securing business of a particular type can be supported by the likelihood of success and the probability of specific profits. These revenue expectations, costs and profit projections need to be thoroughly understood and need to be influenced directly in some way, otherwise they will be controlled by customers or clients. If the costs of doing business are controlled by these groups, however unaware of the situation they may be, then profitability will behave in a randomly distributed way.

One of the key tasks facing marketing staff in any organization is how to bring these aspects under inside control so that sensible decisions can be taken about business development activities and future profit expectations.

4 Strategic planning

Links with marketing

Strategy is essentially about positioning. Its value in business is to ensure the organization is well-placed to take maximum advantage of changes in the business environment.

The military definition of strategy is the advantageous positioning of forces before the action begins. Positioning requires anticipation, and it is useful to look at this positioning as two tasks: planning and execution. The planning activity is often called strategic planning or corporate planning and the execution is in fact marketing, although marketing is seldom defined in this way.

Scope of strategic planning

The concept of strategic planning is not easy to define concisely and it is simpler to examine the subject from the standpoint of its scope.

- Strategy involves positioning an organization with respect to:

  - its markets
  - its operational resources
  - its own management structure
  - its own financial constitution
  - its competition.

This positioning has to be achieved quietly and efficiently whilst maintaining a steady growth in the performance of the organization. In most cases strategic planning starts with setting objectives about the performance expectations of an organization and then assessing the implications for current operations.

Hierarchy of business activities

There is a natural hierarchy of business activities which move from an external focus through to internal action and control. This hierarchy is shown as a simplified model in Fig. 4. Marketing, in these hierarchical terms, forms the bridge between strategic planning and selling.

Marketing and strategic planning cannot be completely separated as the mission for marketing is to implement strategic plans, whether these plans are explicit or not. Organizations that carry out no strategic planning have their positioning determined for them by the activities of competitors, politicians, employees, investors, or customers. Left this way, the positioning that results will have little prospect of providing a secure base for future activities.

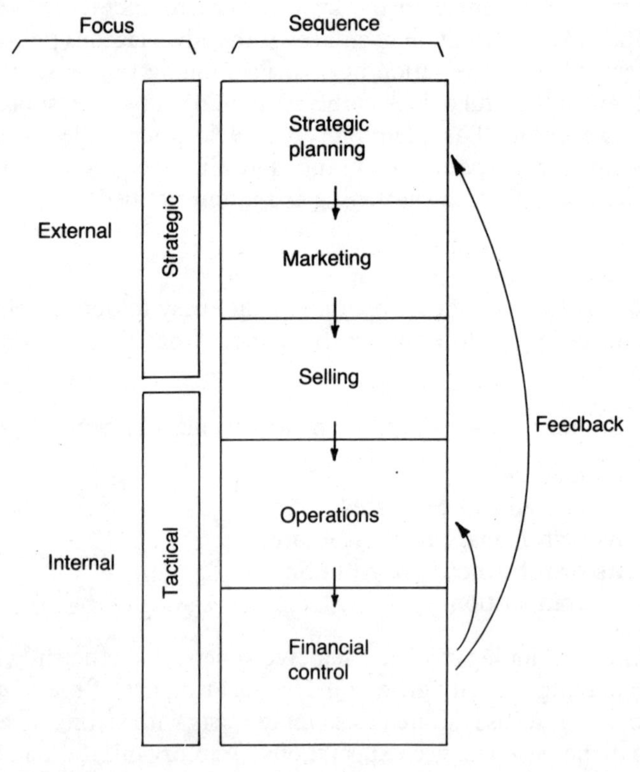

Fig. 4. Hierarchy of business activities

Approach to planning

The starting point for strategic planning is the setting of objectives. There are two general cases to consider here: commercial organizations and public service organizations.

In a commercial organization, investors will demand certain levels of return, which in some way will relate to other investment options that are available. A common way to express these objectives is to target a specific growth in earnings per share. This in turn focuses management attention on profits planning for the future.

Life is not quite so simple in the public service sector. In many organizations there is an absence of any quasi-commercial objectives, although in the UK there has been a movement towards stating performance objectives in terms of a specific return on funds employed. However, the substitutes for commercial objectives can be no less demanding and are often in the form of lists of operational performance parameters, such as costs per thousand households served, etc.

- With the objectives set an organization can address the all-important issue of how these objectives are to be met. This is the critical and creative step in strategic planning and it requires:

  - a vision of what the organization should seek to become;
  - a thorough understanding of market place needs and opportunities;
  - a thorough understanding of the competition that will be faced;
  - an ability to exploit any feature of operations that gives an unassailable competitive edge in the market place.

The outputs from a strategic planning round will be in three forms.

- *Narrative* This will be a statement of the purpose of an organization and its key policies for operation.
- *Targets and budgets* This will be a translation of the planning process into performance projections which usually cover financial aspects.
- *Positioning plans* This will be a description of the ways in which an organization will change shape through its implementation programmes. The most important programmes are usually:

- o marketing
- o organization development
- o acquisitions and divestments
- o technology or product development
- o productivity improvement.

Organizational issues

It is beyond the scope of this book to explore strategic planning in any depth. It is sufficient to say that the process should involve the executive, marketing, and financial functions at the highest level in an organization. Strategic planning can be inadequate for two main reasons. Either the plans themselves are inadequate through faulty assessment or the implementation can be inadequate through bad management.

One factor that influences implementation more than any other is an understanding of organizational issues. Typically, the primary role of the organization is to provide coordination of effort towards achieving predetermined goals and objectives. To effect proper coordination and to maximize the impact of particular skills and resources, a company will adopt a particular style of organization in terms of structure and culture. Any organization is a compromise to some degree and in those businesses which are not marketing-led, particularly if in the fast moving consumer goods sector, there is often a conflict between the coordination needs of the business as a whole and the coordination of marketing effort. However, to address this problem it is first necessary to examine organizational types and company cultures.

Organizational types

Organizational types can be classified simply by considering the degree of centralization and the degree of functional impact. A simple classification scheme is given in Fig. 5.

Centralized operations are typified by an army-style organization. There will be a highly developed command structure and the emphasis is on moving intelligence around the organization for local commanders to interpret against a framework of clear objectives.

Divisionalized operations result typically from the recognition that an organization may be engaged in a variety of activities which

Degree of centralization	Degree of functionalization		
	Centralized (strong)	Divisionalized (dominating)	Decentralized (weak)
Centralized (army)	(11) Autocratic	(12) Ministerial	(13) Tyrannical
Divisionalized (baronial)	(21) Feudal	(22) Matrix	(23) Federation
Decentralized (navy)	(31) Institutional	(32) Mandarin	(33) Privateer

Fig. 5. Organizational types

only partially overlap and which require very different management responses in its different parts. It usually results in the creation of powerful baronies.

Decentralized operations are typified by a navy-style organization. There will be a need for senior people with significant responsibility to operate quite independently in the field without constant reference to the centre, but working within an overall plan.

Increasingly these three simple management structures have been confused by the growing impact of the functional specialist on a business. The organization of these functional contributions can also follow the form of centralized, divisionalized, and decentralized.

The classifications in Fig. 5 have been given simple names for convenience and need some explanation.

- *Type 11: Autocratic* These organizations have few but straightforward business objectives, simple control needs, and a simple management hierarchy. They are usually very well coordinated from all standpoints and often identified by the comment that 'they are quick on their feet'. Unfortunately many companies outgrow this structure as their business activities develop.

- *Type 12: Ministerial* These organizations exhibit their strong functional skills to the outside world and are often seen as quite separate in a divisional sense. There are often few reasons to coordinate activities except to create a collective responsibility for policies and plans.

- *Type 13: Tyrannical* These organizations have few effective checks and balances on the performance of the management team. They can be highly coordinated but many develop questionable objectives or adopt bad business practices.
- *Type 21: Feudal* These organizations have allegiances which respect functional controls, as the operational side of the organization is sometimes incomplete. This organization style seems to be most susceptible to apparently random or irreconcilable commands.
- *Type 22: Matrix* These organizations were once thought to be the ultimate answer for modern business. Whilst communication can sometimes be good, often through a committee structure, the organization tends to become paralysed in terms of action.
- *Type 23: Federation* These organizations are dominated by the divisional barons and are often subject to little or no central control. Communications across divisional boundaries is usually poor or non-existent.
- *Type 31: Institutional* These organizations are dominated by procedures and rules, and are usually unable to handle change effectively. Commercial objectives can be lost completely.
- *Type 32: Mandarin* These organizations create powerful, and often competitive, functional heads. This organizational type is open to political exploitation and communication across boundaries is mostly discouraged.
- *Type 33: Privateer* This hardly qualifies as an organizational type as each component tends to go its own way. At best it is a club, but can be effective from time to time when formal alliances are formed.

Management styles or cultures

To confuse the situation even more, organizations are heavily influenced by the management style or culture that has been allowed to develop. A simple classification of cultural types is given in Fig. 6.

- *Type A: Power-centred* All successful organizations tend to

Power-centred	Merit-centred	Rule-centred	Tribe-centred
● Clear leader ● Clear commands ● Highly focussed ● Vulnerable to: –blind spots –succession –size	● Team approach ● Shares risks and decisions ● Vulnerable to: –domination by one function	● Clear rules ● Stable ● Vulnerable to: –change –commercial pressures	● Benefits employees ● Weak management control ● Vulnerable to: –setting objectives –change

Fig. 6. Management cultures

start out in this form, usually based on the vision or commitment of one person. These organizations tend to be seen as vigorous and entrepreneurial.

- *Type B: Merit-centred* This style probably represents the majority view on ideal organizational cultures. It relies implicitly on the thesis that management is a profession with key specializations. This is often the prevailing culture in public companies.
- *Type C: Rule-centred* These organizations are typical of those where wide accountability is regarded as being paramount even to the detriment of certain aspects of performance. This situation is often found in the public service sector. Ultimate control in these organizations rests with those people who can change the rules.
- *Type D: Tribe-centred* These organizations are run, or more properly allowed to function, by the employee group rather than the management group. This situation is usually the result of a prolonged period of operation in a distressed business environment, and typically demands some measure of community or governmental support to survive.

Coordination

Coordination is such an important issue in marketing and the implementation of strategic plans that a wide understanding of organizational issues is essential. Marketing has to reconcile the 'top-down' pressures of putting a strategy in place and the 'bottom-up' pressures of responding to changes in the marketplace. This can only be done with a centralized coordination for marketing, irrespective of the organizational type and culture of the business as a whole. Marketing organizations, therefore, tend to be quite separate from other line-reporting activities, if they are to be effective. These relationships are never easy to manage. However, without centralized coordination in marketing and without a clear separation of marketing from operations, no marketing effort is likely to produce worthwhile results or initiatives.

Investment planning

Marketing and the concept of investment are closely related. The purpose of investment is to use currently available funds, and forego their immediate benefits, in order to secure a more valuable position in the future. This is essentially the same description as strategic planning and marketing.

Investments will always reflect organizational priorities and organization preferences for a balance or a mix of activities. Priorities should be related to the main positioning objectives that are set and are of primary importance. The question of mix or balance is of secondary importance in the sense that it will reflect a comfort factor for management when contingency planning issues are incorporated into business plans. Mix and balance are concerned with the proportions of:

- new versus old activities
- service oriented versus asset based activities
- contracted revenues versus sales revenues.

Investment plans will be a compromise of pure positioning priorities and the demands of financial prudence or management comfort in an organization. The marketing function should be central to the implementation of investment plans, as it will be the only function with a responsibility to make the plans work outside those functions, such as legal and finance, that make a purely mechanical contribution.

5 Operational planning

Starting a marketing initiative

In many organizations concerned with civil engineering activities there is a feeling that marketing is probably important, that it is probably not done or not done well, and there are conceptual and practical difficulties to overcome in getting a new marketing initiative under way.

There is no classic way to resolve this issue, and this book subscribes to the pragmatic view that:

- an initial budget should be set
- plans should be laid to maximize the benefits of marketing within the constraints of the budget
- the results of the marketing achieved should be reviewed critically and changes should be justified for subsequent budgets and subsequent operations.

Budgeting

- Budgeting is the first practical problem to resolve in marketing, and difficulties are usually encountered in answering questions such as the following.

  - How can the objectives be quantified?
  - How are these objectives related to activities?
  - What resources will these activities require?
  - Are the right levels of resources being provided?

Some guidelines and an empirical approach may be useful in looking at the initial budgeting problem. Marketing needs vary from business to business and Fig. 7 shows some contrasting cases.

Business type	Typical marketing budget	
	As % sales	As % profit
Fast food	5	30
High tech. products	5	20
Consumer durables	3·5	25
Producer durables	2·5	20
Capital goods projects	2	15
Infrastructure projects	1	10
Engineering consultancy	10	30

Fig. 7. Range of marketing budgets

These differing needs reflect practical aspects such as:

- frequency of sales
- price of unit of sale
- focus of purchase decision
- cost structure of product
- purchase criteria.

These terms tend to have a fast moving goods flavour to them, but they are no less relevant to products in civil engineering.

Fast food illustration

- Fast food is characterized by:

- high frequency of sales, say one per week per family
- low unit price, typically paid for out of cash in hand
- simple purchase decisions, usually focused on an individual and usually with a high impulse element
- low material costs in the product, and probably high relative distribution costs
- purchases being the mixed result of availability, price, attractiveness and immediate need.

This whole business area epitomizes marketing for many people and the marketing itself projects an image of brashness, over-sensationalization, and something to look down on. It hardly seems relevant to civil engineering.

For fast food the objectives in marketing are to get the product available for purchasers when they are inclined to buy, and to make the product attractive in terms of price, presentation and quality. These objectives determine the nature of the marketing budget, and the budget reflects the distinctive needs of:

- advertising
- supply to retail outlets
- packaging
- merchandising (i.e. presentation at the point of sale).

Consumer goods illustration

- Consumer goods are characterized by:

- low frequency of sales, say once every two to five years
- significant unit price, typically paid for out of savings or on credit
- simple purchase decisions, usually focused on an individual or a family, but usually determined after some short period of consideration
- relatively high material and labour costs, and probably high implicit costs for servicing in the field
- purchases being the result of product appeal relative to competitive products.

In consumer goods the objectives in marketing are often to provide products that are better at meeting customer needs than those of the competition, and to make these products easily available at an attractive price. The marketing budget for these organizations will reflect the importance of:

- product development
- distributor purchases and stocks
- after-sales support in the field.

Implications for civil engineering activities

Typically, civil engineering activities do not lend themselves immediately to the classic conception of a product, and this is one of the first and one of the major stumbling blocks in marketing. A good starting point is to think in terms of three basic product groups for civil engineering:

- capital goods projects

- infrastructure projects
- professional engineering services.

Capital goods projects
This product group would cover the design and construction of major industrial facilities and process plant. The product comprises:

- Responding to the needs of the market by offering:
  - a design concept which will meet specification
  - a design concept that is distinctive and is based on technical leadership
  - a programme for procurement, construction and commissioning
  - appropriate post-contract or after-sales support.
- Demonstrating credibility to meet the needs of the marketplace in financial, technical and delivery terms.

All these product features have to be marketed, and the marketing budgets will be influenced accordingly.

Infrastructure projects
This product group is similar in many ways to capital goods projects with the main differences being:

- no proprietary processes or technology may be involved
- there is a greater likelihood that infrastructure will involve the public sector rather than the private sector of the economy.

- Both infrastructure projects and capital goods projects are characterized by:

- complicated purchase decision processes with no single purchaser being easy to identify
- long lead times on winning the business
- commitment of significant management and technical resources to prepare an offer
- infrequent opportunities for a sale.

Professional services
It is much easier to conceive a product in this area and much easier to identify clients. This product group usually covers those activities leading up to a client group going ahead with an industrial or process development or developing some infrastructure.

- The product comprises:
  - responding to the needs of the market by offering:
    - an appropriate design concept
    - professional objectivity and integrity in recommending suppliers and solutions.
  - Minimizing the exposure of the client to technical unknowns by adopting a professional approach and by deploying previous relevant experience.

Characteristics of marketing in civil engineering

- Civil engineering would seem to be characterized by:
  - infrequent sales, say once every 15 years up to once only
  - unit prices which are viewed as either capital expenditure or investments, and which need significant attention to funding
  - very complex purchase decisions, involving many people and many different interest factions
  - long purchase lead times
  - purchases being the result of minimizing risks of failure and properly discharging management responsibilities to be objective and accountable.
- In such cases marketing will reflect the importance of:
  - anticipating customer or client requirements
  - deploying relevant experience to provide solutions
  - minimizing customer uncertainties about product performance, cost and delivery.

A word of caution needs to be sounded here regarding the situations set out in Fig. 7. The figures put forward as typical should in no way be regarded as either the right figures nor the norm for the situations described. They are merely representative. Every organization will have its own unique needs and its own distinctive marketing budgets as a consequence. Commercial success and the strength of competition will put limits on the budgets in practice which are not immediately related to the tasks involved in marketing.

Establishing a marketing budget

One of the specific problems in establishing a budget concerns scope. Where for example does one place estimating or preliminary design engineering? All activities which are not generating revenues, but are related to positioning the company to secure revenues, are technically part of the marketing budget. Within this definition will be the sales activity. The simple separation between marketing and selling is that selling starts when a specific opportunity is pursued.

- To set an appropriate budget for marketing the following steps will be useful.

  - Establish a consistent definition for marketing in terms of expense items. Make it as wide as possible even if, initially, not all the expense items are under a single control.
  - Establish what recent marketing costs have been using this new definition and relate them to revenues and profits.
  - Attempt to estimate similar figures for close competitors, or for a division within a large organization, estimate the figures for other divisions. These should be compared and tested against whether they are felt to represent value for money.
  - Carry out an exercise to re-allocate the existing budget in a way which might give better value. It is then possible to examine what budget should be set to meet new targets. Experience in estimating future budgets will improve the whole procedure.

Operational management

The purpose of operational planning is to identify and deploy the resources needed to meet marketing and sales objectives. In many aspects of civil engineering, the very nature of the work makes it difficult to establish what projects result from what marketing or selling activities, and what people influenced the awards, both from the client and the contractor sides. The gestation period on many projects is likely to be measured in years rather than months, and no award is ever the result of the efforts of a single person.

Consequently, parameters that will help in operational planning need to be developed over a number of years, say three, as they can vary enormously from one budget period to another. Again, some guidelines might be useful, as follows.

- Sales targets are the starting point.

- Sales conversion rates need to be established. The simple way to do this is to take the sterling value of successful proposals for a three year period and divide this by the sterling value of all proposals submitted over the same period.

- By applying the conversion rate to the sales target, a submissions target can be established. In each business area, the average size job should be found and this will allow the number of submissions to be made to be determined. This is the basis for all operational planning.
- The sales and marketing effort can now be related to the submissions target so that the number of field initiatives and marketing staff can be established.

Examples of this budgeting process are likely to be misleading, as no really representative figures can be developed. The process described starts with sales targets and ends with the identification of marketing resources based on recent experience. This process is almost certain to produce an unsatisfactory answer first time around in the sense that the resources do not appear to be matched to the sales targets. The response should be to address the performance parameters as a task of management.

For example, if marketing resources are insufficient to meet sales targets on historical performance then management has two options. Either increase the marketing resources if that can be afforded or improve the marketing output from existing resources. The latter case will be the more usual situation.

The first area to improve is the sales conversion rate. This comes from developing a better product and presenting it more effectively.

The second area to improve is the calls conversion rate. This comes from acquiring better intelligence on opportunities for the products and services that are being offered and the willingness of customers and clients to receive proposals.

Both these areas are addressed later in terms of how improvements can be secured. At this point it is sufficient to say that the likely outcome of a budgeting exercise is the setting of targets for sales conversion and calls conversion as a budget balancing item.

Links with selling budgets

Marketing is the act of positioning to secure selling opportunities and selling starts when specific opportunities are pursued. In civil engineering, marketing and selling activities usually involve common resources. There is, however, a vital organizational distinction that must be observed. When selling starts there is an implied commitment to take on a project and the ultimate arbiter for selling inside the organization has to be the person who has profit centre responsibility. This person cannot accept activities sold by other parts of the organization which have been merely passed-on without reference. In recognizing this distinction the usual consequence is an apparent conflict between marketing, which needs to be centrally coordinated, and selling, which needs to be entirely focused on specific profit centres. Budgets, therefore, have to be developed in a way which is centrally coordinated, and then split up into relevant areas of operational and administrative responsibility.

6 Management of intelligence

Basic requirements

Effective marketing requires the effective management of information. Information, or intelligence, provides the oxygen in the system for marketing effort. Intelligence management requires:

- centralized control
- widespread acquisition
- ready availability on demand
- critical examination.

Without centralized control of marketing intelligence, irrespective of the structure of the organization, there can be no effective way of coordinating the efforts of a company.

Categorization

- Intelligence has to be gathered from all meaningful sources and processed. In general, intelligence comes in two forms.

- *Hard intelligence* This comes from published information in reputable journals or gazettes. This intelligence is typically widely available and gives little opportunity other than to respond to invitations. Organizing hard intelligence is a relatively easy process, but it is often badly neglected.
- *Soft intelligence* This comes from field reports, business meetings and conversations. It is more difficult, often, to verify this information and it is usually very perishable. Properly managed, soft intelligence is the key to outstanding marketing.

Scope

Marketing intelligence for civil engineering needs to cover the following aspects.

- Country-by-country economic and political projections

  - It is essential to be well-briefed regarding the key forces driving developments around the world. A well-briefed representative has a sharper perspective on opportunities and key personalities.
  - Marketing effectiveness often comes from a sound choice of priorities and priorities in civil engineering often have a strong geographical dimension.
  - In many instances, it will be necessary to look at countries on a regional or provincial basis before satisfactory conclusions can be drawn.

- Sector-by-sector analyses

  - Most organizations will specialize in certain areas of activity. The world markets for these activities and the customers or clients that make up these markets need to be thoroughly researched and understood.
  - Attempts should be made to establish what the future plans of key clients are in key markets around the world.

- Technology trends

  - Within each sector it is important to assess trends in technology and product or service requirements.
  - This is essential if positions of leadership are to be protected.

- Activities of possible associates

  - Comprehensive details need to be kept on organizations that can be involved in the business process as working partners or as commercial partners.
  - In civil engineering these details should cover:
    - consultants
    - contractors
    - financing institutions.

- Performance of competitors

  - It is important to understand the business operations of competitors in as much detail as internal operations can be understood.
  - Competitive intelligence should cover:

- sales awards
- profitability analyses
- distinctive products, leadership positions, customers.

Control

It is essential to have the management of intelligence under centralized control and have it closely integrated with marketing management.

- Intelligence is an expensive commodity and is very susceptible to:

- o duplication
- o difficulties in establishing independence and origins
- o being perishable
- o difficulties in assessment or interpretation
- o indiscriminate mixing of facts and opinion.

Intelligence therefore demands management by people not only knowledgeable of the organization and its activities, but in addition, people skilled in the management of information.

Acquisition

New techniques of information acquisition and retrieval, which have become increasingly powerful since the early 1980s, are transforming this area of activity. However, a note of caution is appropriate. Advanced retrieval capabilities now outperform acquisition capabilities, particularly in the field of soft intelligence text. No system is better than the quality of its acquisition capability.

- Key sources of information are:

- o published materials
  - news
  - special reports
- o field reports on day-to-day activities
- o special field research interviews
  - customers and clients
  - associates
  - trade associates, embassies, etc.

It is not sufficient to rely on published materials alone. This will not give adequate competitive advantage. Field reports are a neglected area in many organizations. Those people in daily contact with the market will have some of the sharpest perspectives on what is happening. These perspectives can often be very uncomfortable as people in the field very quickly identify with customers and clients and tend to get detached from their own organizations. Their reports are often critical implicitly of the performance of their own organizations. However, with proper structuring these field contacts can be turned into rich veins of intelligence about trends, concerns of the market, and the competition.

The critical factor in making field reports effective is the process of central acquisition and assessment. This has to be carried out by senior people with a deep and respected understanding of business operations in the organization.

Special field research is usually regarded as an occasional need in most organizations. However, the most effective organizations regard this as an essential ongoing need and justify the cost of having it carried out objectively by skilled professionals.

Dissemination

The ultimate purpose of intelligence is to generate action. However, all information needs critical examination before its release, and this interpretation process is typically a much-neglected subject.

It is remarkable how the same set of data can lead to different conclusions being drawn by different people. Data or information should never be presented without conclusions, and the data should support the conclusions not lead them. It is the responsibility of intelligence management to establish these conclusions and to do this with assistance from other management functions as necessary. Conclusions cannot be developed in isolation.

The process of dissemination of intelligence is the one that makes marketing come alive. When it is done well it gives strong signals to the outside world about being sharp and well-coordinated, and gives strong signals inside the organization about being on top of events.

- The format for information dissemination in an organization should be a short report document which states clearly:
  - the subject or issue

- o the conclusions drawn
- o the authority for the views.

These reports will become central to marketing within an organization and they need careful filing for future referencing.

Interactive needs

Ready availability of information on demand is of prime practical importance. However, this can cut across the principle of never allowing information to be passed without first understanding the demand, then undertaking responsibility to draw a conclusion, and finally presenting the information in a way that will aid action or decision-making. The dilemma can be resolved by adopting the following practice.

Before a decision is taken or a course of action committed, there should be a briefing session which draws on all the available intelligence. Once action has been agreed and implemented there should be a corresponding debriefing session. This simple discipline, when rigorously applied, will pay significant dividends.

Summary of priorities

The main priorities to tackle in managing intelligence will be:

- Keep good records of past experience; these records should be cross-referenced by type of work, client, value, location, associates.
- Invest in a first rate library facility with the capability to follow through on all items of technical and commercial interest
- Ensure that all published information relating to invitations to bid for work is picked up and disseminated to those who should respond.

7 Promotional activities

Introduction

Promotional activities concern customer–marketer relations. They have nothing to do with junketing; they are essentially concerned about managing different levels of awareness in customers and clients. An outline of customer–marketer relations is given schematically in Fig. 8.

Customers and clients have to be taken from a position of being totally unaware of the activities and relevance of the organization, right through to being a contented purchaser of products or services.

- The process involves:
  - projecting an image
  - creating awareness
  - making prospects knowledgeable
  - establishing conviction amongst prospective clients
  - turning prospects into clients and customers
  - maintaining commercial loyalty.

Image

No review of marketing is complete without commenting on the subject of image. All organizations create an image and many organizations project more than one image, when different recipients are taken into account.

Marketing has an overall responsibility to create and project a suitable and appropriate image for an organization. Image as a concept is not at all as intangible as it first appears. The real problems lie in creating an image which is truthful, appealing and unique, and consistent with existing reputation.

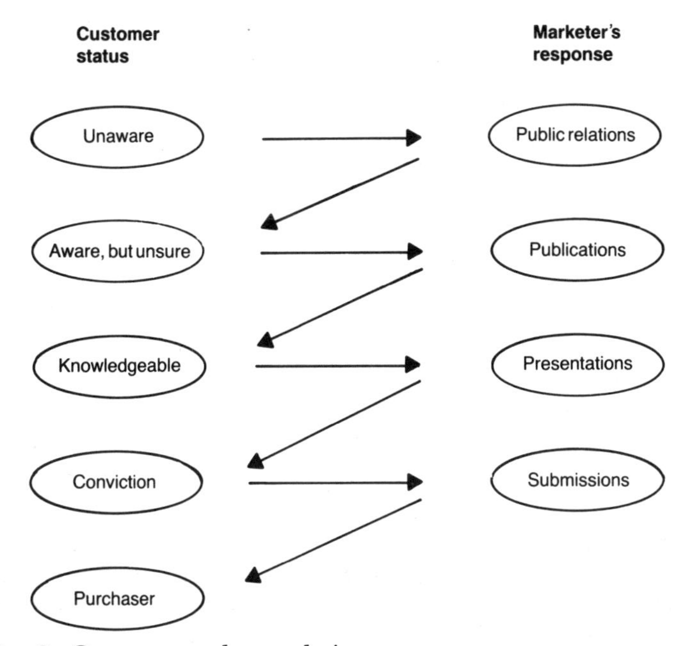

Fig. 8. Customer-marketer relations

Creating an image starts with the communication of a corporate identity. This identity is created from physical components such as signs, products, stationery, advertisements, brochures, offices and office locations. The identity is a factual statement about an organization in non-language terms and is essentially truthful. The style of presentation and the language statements that are made to support it create an image in the eye of the recipient. This image can be far removed from reality when the presentational style, the projection and the language used to support it are inappropriate.

A new profession has grown up around the need to create corporate identities and image. All organizations need to attend to image if they are to compete in world markets; it cannot be left to chance. A suitable projection should create truthful impression of:

- the scope of activities
- the management style of the organization
- the successes of the organization
- the standing of the organization in the business community.

False projections are remarkably easy to detect, yet truly appropriate ones are not easy to articulate although are easily recognized when they are right.

Creating a corporate image is beyond the scope of this book but it is instructive and rewarding to try to identify a popular car that best creates the general image of an organization and another that corresponds to the image it desires to create.

House style

Marketers should pay attention to house style and the disciplines needed to have it applied consistently. House style is part of the corporate identity statement and a powerful aid to marketing.

- The first items to attend to are:

- use of corporate logo, with local variants to be discontinued
- consistency in stationery, business cards, compliment slips, again with a ban on unauthorized variants
- consistency in all sign boards, with instructions made to keep them clean.

A well presented and consistent house style creates an impression of efficiency and attention to detail, whereas wide variants in style indicate divisions or confusion in objectives. It is interesting to note at this point that house style is not a camouflage for reality, as the process of getting agreement in an organization on house style is itself a powerful force for resolving any confusion in objectives.

Public relations

The mission of PR is to tackle the basic levels of awareness amongst customers, prospective customers, and the general public with respect to the activities of an organization. Improved understanding and increased awareness of activities will come from PR initiatives which focus on:

- developing business contacts at a personal level
- developing media understanding of activities
- publication of materials to support activities
- general advertising where relevant.

Whenever possible, PR should be seen as the mechanism to spread awareness as a preliminary to future business development. Too often PR is used defensively to explain away difficult situations or poor service aspects.

The most powerful PR platform is an enjoyable social setting as it allows business relationships to develop along another dimension. The objectives here must be twofold:

- to ensure that there is a proper awareness on both sides about respective activities
- to reduce the barriers to future contact when business opportunities arise.

The whole area of PR is often mistrusted and misunderstood which, ironically, implies that PR work on PR departments has been sadly deficient. The work needs to be carried out by a senior person in every organization and not be treated entirely as some mechanical chore. The reason for this is that PR is often cast in the role of organization spokesman, and that spokesman must be sufficiently senior to express views and take positions without constant referrals to other executives.

The role of the media

The primary interest of the media is news. All future business developments and possibilities are news and this often generates public attention at times when private negotiations are delicately poised. There will always be tremendous internal pressures to keep business development aspects that are still unresolved as confidential as possible. Being uncooperative with the media about this does not augur well for those occasions when media help is sought to promote a particular point of view on topical business issues.

- There is only one satisfactory way to approach this area, which is as follows.

- Accept the need for the media to chase news items hard, and expect them to present it entirely in their own way.
- Create a position of respect and trust within the media community, so that when sensitive news breaks, a favour can be called on delaying copy in return for an up-to-date briefing on the situation involved.
- Create the position of trust and respect over a period of time by:

- offering regular and objective briefings on activities without any attached expectations on any reporting favours
- taking care to ensure that the level of business understanding in the media about the organization and its operations is first rate
- being available and willing to comment on a wide range of issues when asked.

Information and comment is the trade currency of media relations, and it is a genuine two-way street once suitable relations are established. When this aspect of PR is properly managed the benefits are:

- dynamic contact with current business issues and opinion
- the implicit projection of the organization as an industry leader
- the ability to contain sensitive issues until they are ready to expose.

Publications

- Corporate publications fall naturally into three categories:

- report and accounts
- brochures
- professional articles.

Statutory requirements will usually demand the publication of annual reports and accounts. These publications are expensive, even when the production itself is modest in scope, as a consequence of the compilation and mailing costs. The annual report and accounts in many organizations represent a wasted opportunity. With some care and attention it can be a first-rate production carrying a strong message about the scope and success of current activities. It is the single most important publication in any organization and should be well written, well illustrated, tastefully presented in a consistent style, and be made widely available.

Brochures are a feature of modern business activities and most organizations are expected to be able to produce brochures on demand. They are seen by many people as an important aid to selling. Undoubtedly, some brochures are needed to support business operations, but in general there is a tendency toward:

- o too many brochures, which then cause confusion about relationships within a business
- o too many different styles of publication, which again creates confusion about ownership, objectives and relationships
- o a lack of discipline in keeping publications current and, more importantly, in withdrawing old and inappropriate publications.

Although brochures are an aid to sales, and may well be only used by specific parts of an organization, overall control needs to be exercised by a central marketing function.

Writing professional articles offers a powerful and cost effective way of increasing awareness in the business community. Article writing should be encouraged throughout an organization with the objective of placing material in prestigious journals and magazines. These articles are in fact rarely read by the people one would like to read them, so the follow-up activities in this exercise are vital. All articles should be obtained as reprints (most magazines are only too happy to provide this service at a modest cost) and they should then be mailed to the people who should have read them. This process will associate key staff in the organization with a prestigious journal and with their taking a role as an opinion-former in the eyes of prospective clients.

Advertising

Advertising is very expensive. It is most effective when it can get direct access to ultimate consumers or end-users. In civil engineering, the ultimate consumer is very hard to identify and is not likely to exhibit the persona of an individual. Mass communication techniques, such as advertising, have to be addressed therefore in a distinctive way.

- Advertising is used typically to meet the following four objectives:

- o to communicate information about products, services, and operations
- o to support and help shape a corporate image
- o to provide constant reminders about the presence of an organization

- o to induce prospective customers to purchase products or services.

- • The problems in establishing the relevance of advertising in civil engineering focuses on three questions.

- o What audiences need to be influenced and why?
- o What messages should be developed?
- o How can results and effectiveness be measured?

Audiences

- • All organizations need to communicate with four audience groups:

- o prospective customers
- o the financial community
- o the general public
- o employees.

Prospective customers need to be made aware of the products, services and business stature of an organization. The mission is to encourage business transactions and create a preference in favour of ones own organization. A major problem in civil engineering is identifying the customer.

The financial community need to be kept well informed about results, developments and the general scope of activities. Without this understanding, the business stature and rating of an organization could suffer significantly. The financial community has a strong interest in civil engineering organizations as the whole area is seen as an important indicator of future economic health.

The general public play an important part in influencing the customer in civil engineering. There is a strong commitment to public accountability, either implicitly or explicitly, in civil engineering work. The general public need to be knowledgeable and favourably disposed in general to organizations involved in construction activities.

Employees are often a forgotten audience for advertising. They have a greater general interest in any material than the general public and will be critical of the objectives, messages and style of the advertisements. They will look for images that ring true and which command their support, and they will look for messages they recognize and subscribe to. They are, however, a secondary audience in the sense that advertising aimed at employees alone is rarely justified.

Campaigns

- The prerequisites of an advertising campaign are:
  - a targeted audience
  - a simple objective
  - a focused message
  - an appropriate selection of medium
  - an adequate budget
  - a method of measuring results and the effectiveness of the campaign.
- Campaigns, in general, fall into two groups:
  - corporate advertising
  - product or services advertising.

Corporate advertising should be used to establish a position and standing in the business community. All organizations need some form of corporate advertising.

Product or services advertising should be directed at specific prospective customer groups. If no groups are easily defined, there may be no case for any product advertising. In civil engineering, in general, this form of advertising has questionable value.

The whole advertising area is very highly structured in terms of available professional services and agencies. Any advertising campaign should be undertaken in conjunction with a relevant agency. It will probably be an agency specializing in business-to-business communications. One note of caution here. First contacts with advertising agencies can be a culture shock for the uninitiated. However, the biggest issue to get to terms with is the way in which the advertising business charges for advice and services.

Budgets

It is extremely difficult to establish what budgets should be allocated, if any, for advertising and for what types of advertising. There is no valid or accepted way of approaching this issue. Most advertising agencies will repeat an old American truism which says '50% of all advertising is wasted but it is impossible to say which 50%'.

A practical way around this problem is simply to set aside an acceptable sum of money and merely monitor whether in qualitative terms a campaign was perceived to have given value for money.

Research

Some quantification is possible in advertising. Provided proper preparations are made, it is possible to measure shifts in awareness and knowledge over time as a result of advertising campaigns. Before any campaign starts a research programme should be undertaken to meet the following objectives:

- to establish what the prevailing perception of an organization is amongst the target audience
- to establish where perceptions need to be changed and the extent of the change needed to meet the marketplace projection that is being sought.

This research step provides two benefits. It establishes a datum to measure subsequent views against and it points up the key messages that need to be communicated if false perceptions are to be corrected.

Secondary advertising

Most organizations engage in secondary advertising, such as advertising for recruitment purposes, and opportunities will exist to use such programmes to support wider corporate objectives. A great deal of recruitment advertising is read by people who have no immediate interest, but may in some way be influential in other fields of endeavour. Recruitment advertising communicates an image, which needs to be related consciously to the desired corporate image of the organization, and it signals immediate development priorities when key skills are being sought. This whole area needs to be brought into the overall view about corporate projection.

Summary of priorities

The main priorities to tackle in establishing new promotional activities will be:

- Prepare a first rate annual report and accounts with good text and relevant illustrations;

- Encourage key staff to write articles on subjects of topical interest in prestigious publications;
- Rationalize and clarify the existing house-style and ensure that it is consistently applied throughout the organization.

8 Presentations

Objectives

Presentations which are staged face-to-face events between proposal teams and prospective clients are the highpoint of a great deal of marketing effort. All previous activities in marketing have been directed towards creating the opportunity to meet people with a need. Looking at presentations from this standpoint their real cost can be quite staggering. Yet, the whole area is often a Cinderella activity in civil engineering.

The purpose of presentations is to communicate with interested parties in a structured but reactive way. Presentations should be used for the following tasks in marketing:

- o to inform various audiences and centres of influence about operations and experience
- o to promote specific proposals to work with prospective clients.

Legal profession experience

The legal profession can claim six hundred years of accumulated experience in presenting well-constructed points of view. This experience is relevant to marketing.

- It requires the following disciplines:

- o the rigid adherence to a presentational format
- o the commitment to develop a single thesis for all presentations
- o a working knowledge of the rules of evidence
- o an understanding of what properly constitutes proof.

These points at first sight may seem far removed from either marketing or presentations, but they are in fact central to all exercises in communication.

Importance of preparation

The preparation needed for presentations is extensive. The first objective is to establish the message that needs to be communicated. All messages should ultimately be framed in the form of recommendations for action, and this should be the mission of all presentations.

These recommendations will need careful development and will need to have supporting evidence and reasoning when put forward. The process of developing recommendations is classically:

- assembly of evidence and its categorization
- development of conclusions and their testing for validity
- assessment of implications on preferred courses of action
- development of recommendations.

This process comes quite naturally to scientists and engineers although the process is still susceptible to:

- application of false logic in assembling evidence
- faulty conclusions
- inability to define action requirements.

One of the biggest failings is the inability to conceive or recognize conclusions. They are in fact abstract statements which form a bridge between evidence and action, and serve to justify the action taken. This issue will be dealt with more fully.

The secret in preparing for presentations is to realize, drawing on the experience of the legal profession, that the sequence of presentation is exactly the reverse of that to develop the material for the recommendations, then to relate these to key conclusions, and then to use only that evidence needed to support the conclusions drawn. A good presentation is not an excursion through the diary of development.

Critical role of conclusions

What is so special about conclusions? Consider the case of the Channel Tunnel. There were clear signals from the authorities that the Channel Tunnel would receive political support if it could be funded entirely from the private sector without any form of Government guarantee. As the project was large, by any measure, it was possible to conclude that:

- o the least cost solution would be very attractive, even if this did not provide the best prospective return;
- o using well tried technology would give shareholders and lenders an added degree of security;
- o reducing the scope of the project to the absolute minimum consistent with meeting the broad traffic opportunity would also make shareholders and lenders more comfortable.

This should have led to the immediate realization on the part of the bidders that a simple rail-only tunnel was the most attractive solution. It also turned out to be the least costly and to have the best return as a result of geology and shipping considerations. However, a number of bids were proposed which attempted to provide directly for road traffic, as this was what the general public wanted, even though this added extra cost and involved new and untried technology.

In this example, drawing the right conclusions within the Channel Tunnel Group about the probable attractiveness of the solutions led to a very strong competitive position. In contrast, drawing false conclusions, such as attempting to give the general public what they wanted when they were not the group putting up the investment, led to untenable positions.

Conclusions are easy in retrospect, but those organizations capable of developing their forward business based on drawing outstanding conclusions about needs and opportunities are going to be tomorrow's leaders.

General format for presentations

Presentations need both a structure and some form of linkage to move from one aspect to another. At the outset it is important to appreciate that the structure and the linkage should be independent of the subject matter. This means that the structure and linkage are common to all presentations.

The general format for presentations proposed in this book is given in Fig. 9.

- The first step is to establish clearly the subject or issue to be presented. This is often still necessary even when introductions have been made by the chairman of the session.
- The second step is to effect any outstanding introductions

Sequence	Linkage activities	Message structure
1		Subject/title
2	Presentation outline	
3	Definitions of key terms	
4	Objectives of presentation	
5		Recommendations
6		Conclusions, supporting recommendations
7		Evidence
8	Summary	
9	Any questions	
10	Courtesies and wind-up	

Fig. 9. Presentational format

relating to the presentation team and then give the audience an outline of the presentation. This step is vital as it gives people a routemap and gives them greater certainty about where issues will be raised.

- The third step is to clear away any special terms or definitions which may be used and which are essential to the understanding of the material. Often this step is not necessary.
- The fourth step is to restate the subject matter in the form of objectives for the presentation. If people are going to be asked for a decision, let them know at this point. The objectives should reflect the objectives of the presenters alone.

At this point the audience is positioned and prepared to receive the contents of the message contained in the presentation.

- Step five should be a clear statement of what recommendations for action are being proposed. This will stop the audience worrying about where the presentation is going. It will also give them a frame of reference to assess the reasoning and evidence.

- The sixth step is the most important part of the presentation. It should be a statement of conclusions in the form of a simple and attractive thesis. This is the step that will differentiate this proposal from the competition. It will allow the quality of the proposals to be demonstrated and related to the strengths of the organization. When competitors fail in this step they afford others the opportunity to set the checklist on which they will be judged.

- The seventh step is to present the relevant evidence to support the thesis presented. All evidence should be in the form of a single best reason followed by all the arguments that can be mustered against the thesis, together with an adequate answer which renders them invalid.

- The eighth step is to provide a summary of the recommendations and conclusions. This should be the final part of the formally structured material and should end with some simple statements about what should happen next and why.

- The ninth step is to open up the presentation to questions. This question and answer session also needs some structuring if it is to be successful.

  When a question is asked, it should be rephrased by the presenter and repeated to the audience, for the following reasons.

  - Not all of the audience may have heard the question.
  - The process allows the presenter to break contact with the person asking the question and also draws in the rest of the audience.
  - It gives some thinking time, although this may not be necessary.

Answers should as far as possible relate only to issues that support the conclusions and recommendations in the main body of the presentational material. The two main traps with questions are:

- o to present answers which are in conflict with the thesis advanced
- o to introduce a whole new subject area into the debate which may not be directly relevant to the issues being discussed.

- The tenth and final step is to wind-up the question and answer session, although often the chairman will do this, and then thank the chairman, the hosts, and the audience in an appropriate way.

Managing the props

Presentations are highly technical affairs and are becoming increasingly more sophisticated. In general, the best practice is to keep things as simple as possible. However, three aspects of presentations need careful management and staging:

- o the room or auditorium
- o the equipment
- o the presentation material.

The venue

- All rooms should be checked out in advance with sufficient time to make alterations if necessary. It is important to:

- o ensure that the audience can hear
- o ensure they can have an uninterrupted view of any projected materials, and that the materials can be read
- o arrange the speaking position so that it is convenient for managing the presentation materials
- o occupy positions which allow fairly free movement in the audience for people coming and going.

Choosing a medium

- There are few main options in choosing the presentational medium:

- o delivery from a lecturn, with or without notes
- o speaking to overhead projector slides

- o speaking to 35mm projector slides
- o working with a flip chart.

Speaking from a lecturn with no props is probably the most difficult. It demands a first rate delivery which is capable of not only holding the attention of the audience, but also of being able to communicate complex arguments and quantitative evidence. This mode of presentation is best left to those situations when the audience has turned up to listen to the presenter as a personality and not necessarily to listen to what he has to say!

The other three options are much more suited to most situations as they give the speaker some support and they give the audience some extra interest. The choice is a matter of mechanical preference and Fig. 10 outlines some of the advantages and disadvantages of each. Possibly overhead slides are the best single choice as they represent a good compromise between:

- o simple equipment and auditorium needs
- o simple materials preparation
- o allowing eye contact with audience to be maintained
- o flexibility in sequencing at short notice.

Managing the equipment

- There is no excuse for equipment failure. Before any presentation check that all the equipment is working properly. Pay particular attention to:

- o power cables, which are susceptible to being kicked out as the audience assembles
- o projector bulbs
- o flip chart markers.

Never make an important presentation without having a spare projector bulb on hand and without knowing how a failed bulb can be changed. If the equipment being used is provided by others then enlist appropriate help if necessary.

There is little more disorienting and disturbing for presenters and audiences than to be faced with an initial equipment failure.

The presentation materials

- All presentational materials should of course be simple and

	Advantages	Disadvantages
Overhead slides	● Can be shown without blackout in auditorium ● Presentational material is always between the presenter and the audience and can be used as a checklist during presentations ● Very flexible if material has to be re-sequenced to answer questions.	● Requires a lot of work to select, position and stack slides during a presentation ● Involves bulky equipment which is difficult if equipment has to be taken to a venue
35 mm slides	● Quality of projected image ● Convenient click-on and click-back facilties ● Convenient to transport	● Requires a dark room – not always easy – breaks audience eye contact ● Susceptible to sequencing problems
Flip chart	● Requires minimum preparatory work ● Allows development of material during presentation	● Can only serve a limited audience ● Mechanically difficult to use and speak with ease ● Sometimes regarded as amateur by the audience

Fig. 10. Medium options for presentations

clear to read. However, the biggest traps in preparing materials are:

- o trying to make too many points on one exhibit
- o cluttering the exhibit with too many words
- o using exhibits to project sentences which are then read out to the audience.

Each exhibit can be drafted according to a strict format. In general there are two types:

- o formats to set out agenda or to link one issue with another – these are simple checklists which are needed to help the audience and the presenter establishing where they are within the presentation;
- o formats to communicate recommendations, conclusions and evidence – a suggested format is given in Fig. 11 which allows

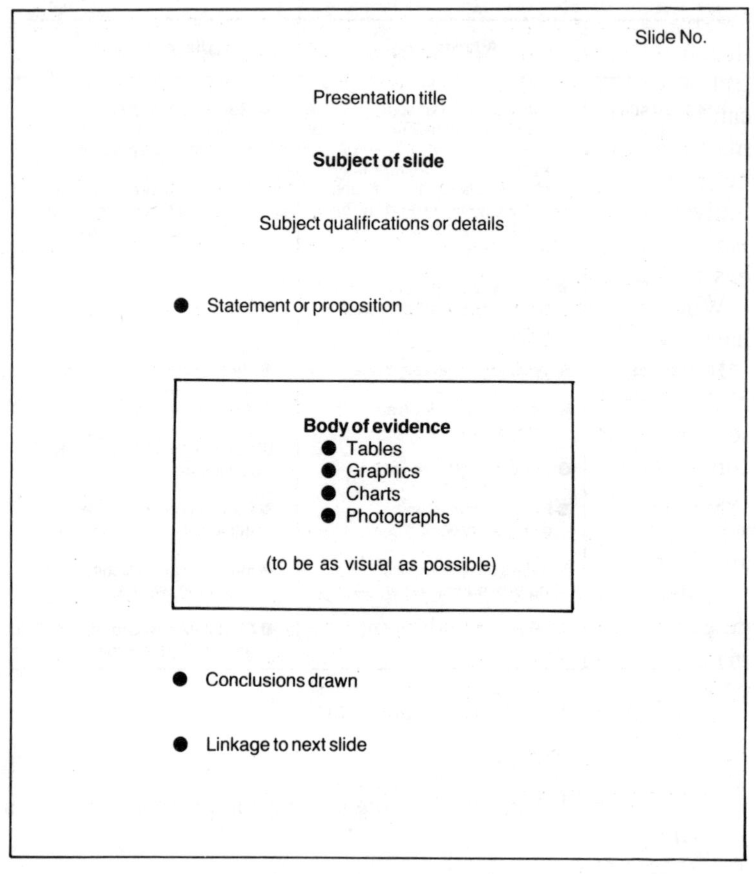

Fig. 11. General format for presentation slides

a presentation to move from one point to another with sound logic and good linkage.

Although the format is rigid, it is flexible in use and is not dependent on subject. It is relevant for all situations.

Managing the audience

There is only one weapon a presenter has to manage his audience: eye-contact. It has to be mastered, as everything else is secondary.

It is essential to establish eye-contact with all parts of the audience during a presentation. It should be done more than once. It requires looking directly at people for between five and ten seconds. Less than five seconds creates a sweeping-effect which make the audience feel they are being given scant recognition. More than ten seconds can lead either to embarrassment in the sense that an individual may be pounced on and a response requested, or a response may be generated when none was sought. Repeated returns to one individual or one group create the impression of favouring or pandering to some sectional interest in the audience. Eye contact has to be made democratically.

When it is well done, an audience will feel that they as individuals have had the benefit of the presentation.

If good eye contact has been established during a presentation it is a powerful control weapon during a question and answer session. Breaking eye-contact signals the end of a brief individual contact and the opening up of the issue to the audience in general. A common failing in presentations is to have no eye-contact whatsoever with the audience when formal material is being presented and to follow this with exclusive eye-contact with questions.

In general it is important to realize that the people in an audience deserve consideration. Avoid monotonous and heavy sessions when planning a presentation. If possible walk around during a performance and make the most of the visual aids.

Performance aspects

- There are two aspects to performance in making presentations:

  - technique
  - appearance.

Technique has to be developed with practice. Skills need to be acquired in:

- memorizing the outline of the material being presented
- remembering to make relevant introductions and to thank people for the opportunity to make a presentation to them
- projecting a clear voice
- being confident enough to break entirely from the presentational material to establish contact with the audience and to be sure of returning to the material at the right point
- projecting something of one's own personality.

When presenting, always stand up straight with hands down the side and the jacket buttoned-up. This posture is often regarded as uncomfortable and unnatural. However, it looks superb from the audience's standpoint, and it provides few distractions or irritating behavioural habits. In fact, it is neither uncomfortable nor unnatural. The first experience of it gives this impression only because being fully exposed during presentations is not a common experience.

Appearance when presenting is vital. Be well groomed, generally conventional, and with nothing to indicate membership or support of any kind of affinity group. Appearance for presentations should be businesslike in the context of the presentation itself.

Summary and golden rules

When making presentations, the golden rules are:

- have all materials prepared professionally
- keep all materials simple, with as few words as possible
- organize a dry-run and full dress rehearsal beforehand
- check out the facilities and the room before any presentations
- when presenting, stand up straight with hands down the side
- develop eye-contact with the audience
- present a conventional appearance
- structure all material to lead to a conclusion
- end the presentation with the conclusions and recommendations and sit down. The chairman will handle the rest of the business
- have a game plan devised for answering any questions
- do not behave as if you would rather be somewhere else, no matter how uncomfortable the circumstances.

The full dress rehearsal is essential. This can be a most humbling experience for even the most experience people, but it is always rewarding.

9 Submissions

Objectives

The development of proposals is another critical activity in marketing. A proposal is an offer to provide a service to a client. The proposal document is often referred to as a submission and it has contractual implications. Typically it comprises a project outline with supporting details on finance, resources, programming and costs. A proposal submission will communicate, whether intended or not, many comments on the style and competence of the organization. Submissions should set out to meet the following objectives:

- to demonstrate a clear understanding of the need that is to be satisfied;
- to present specific proposals in a clear and simple fashion;
- to demonstrate the competence of the organization to take on the tasks involved;
- to focus on the single most important reason for being selected by a prospective client.

The use of proposals

- Proposal submissions are usually multi-functional documents. They are the medium for:

- making a specific offer
- outlining contractual terms
- specifying how results will be achieved
- establishing a case for selection.

They are therefore documents which need both to generate action or precipitate a decision, and to fulfil the role of a reference work if things proceed. These two quite separate aspects need quite separate attention.

Preparation as part of the research process

The preparation of documents is an essential part of the research process. Documents should never be developed at the last minute after all the decisions or options have been determined. This merely produces a poor document and misses the opportunity to use the contribution that its preparation can make to picking the right options. In fact, the preparation of the document is the proposal itself and should start on day one.

Bid strategies

- There are certain key decisions to take when responding to an opportunity to bid for work.

  - Is it worth submitting a proposal?
  - If it is, how can it be differentiated from the competition?
  - What posture should be taken on pricing?

When to bid

- Before a submission is developed the following factors should be checked:

  - that there is a real prospect of the work going ahead – many invitations to submit are driven by the need to establish some aspect of feasibility on the part of the client;
  - that adequate contacts exist with the customer or client already – being completely unknown at a personal level is a great disadvantage;
  - that the prospective client regards your organization as one of the leading two or three contenders;
  - that the organization is able to commit proper time and resources to the preparation of a bid;
  - that it is possible to see areas of strategic or competitive advantage that the organization can offer;
  - that a preliminary numerical analysis indicates that adequate returns can be made.

Evaluating an opportunity to bid

Some organizations, particularly those working in highly technical public sector areas, adopt a scoring process to evaluate the attractions of submitting a bid. Bid costs are often significant items reflecting time involved and support resources needed. When there is a high cost in preparation it is important to get the bid priorities right.

- The factors to score and evaluate are:

- o relevance of past experience
- o technical capability
- o team strength
- o support facilities
- o marketplace knowledge
- o customer contacts
- o competition
- o opportunities to meet expectations
- o prospective returns
- o relevance to future developments.

A scoring scheme to do this is given in Fig. 12. Unless a score of 75 or more can be achieved it is unlikely that a bid would succeed even if one was prepared.

Pricing

In straightforward bid situations it is very difficult for work to be awarded to anyone other than the lowest bidder. If the prospective client is a public sector body and has taken great care in selecting a short list of bidders, then it is even more difficult to make an adequate case against the lowest bidder. One can only charge a premium for superior technology, experience or management. The bid evaluation process should help to form views about how realistic it is to put in a premium.

Sequence of preparation

A suggested sequence of preparation for submissions is given in Fig. 13. It comprises ten main steps.

- The first step is to carry out some background analysis. This should have three aspects:

- o an analysis of customer or client requirements
- o an analysis of likely competitors
- o an analysis of why a submission should be developed.

Factor	Scores			Rating
	Positive	Neutral	Negative	
	10 9 8	7 6 5 4	3 2 1	
1. Relevance of past experience	● In-house	● Subcontracted	● New area	
2. Technical capability	● Proprietary leadership position	● Capable	● Needs development	
3. Team strength	● Best in field	● Best available	● Second team	
4. Support facilities	● Available	● Subcontracted	● Needs development	
5. Marketplace knowledge	● Active contacts	● Aware	● Did not expect invitation to bid	
6. Customer contacts	● Good working contacts	● Occasional	● None	
7. Competition	● Limited	● Open to wide number	● Obvious likely winner	
8. Customer expectations	● Can be exceeded	● Could be met	● Difficult to meet	
9. Commercial	● Can charge a premium	● Standard margin	● Needs cost cutting	
10. Relevance to future	● Supports a leadership position	● One of many such opportunities	● One-off situation	
			Totals	

Fig. 12. Scoring bid attractiveness

- Having decided to prepare a bid, the next step is to compose a strategy designed to win the bid. This is a statement that needs to be written down, needs to be convincing, and needs the support and commitment of the organization. It needs to cover:

  - how customer liaison should be established
  - what main feature should be promoted as the single most important issue in securing the customer's commitment

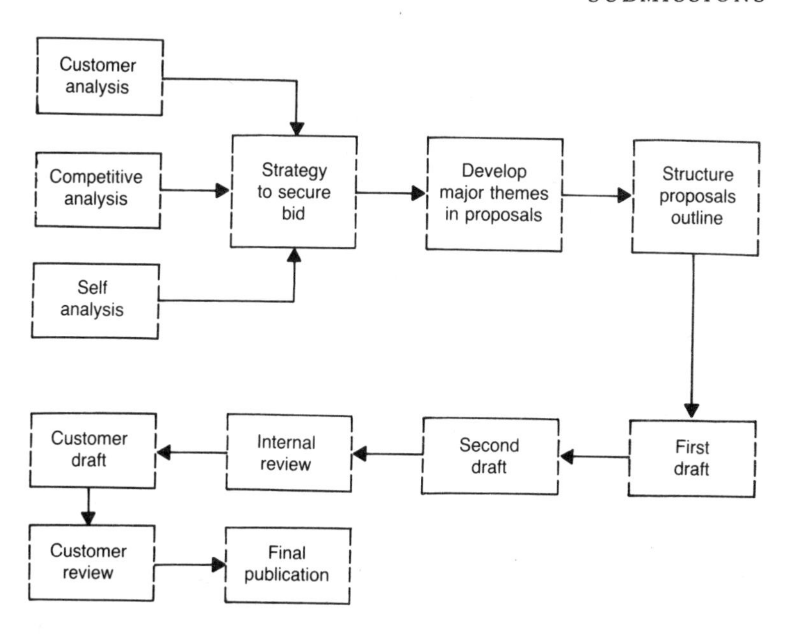

Fig. 13. Proposals preparation

- o what priority on resources the bid preparations should command.

- The third step is the major research step where the arguments, themes and approach options are developed. It is useful to draw this step to a close by attempting to present the main themes for the document in the form of a scenario. This step, although vital, should be accomplished relatively quickly. It is, after all, mainstream to the competence of the organization and can be developed from experience and marketplace knowledge. If some of these themes need later refinements the rest of the document development sequence will provide ample opportunity to do this.

- The fourth step is to produce a document outline and timetable. Great care should be taken at this stage as the plan and outline should not change at all during the preparation. There are a number of factors to consider in structuring an outline. The main sections should be:

- o an executive summary
- o a technical section
- o a management section
- o a commercial section.

Within each section, care should be taken to structure the material so that it is clear and easy to evaluate by the client. To do this requires some research but almost all clients are more than happy to discuss what is important to them and what relative values they place on certain technical and management features.

The technical and management sections should be written or developed first and these should be followed by the commercial section. The executive summary should be prepared last.

Suggested formats for submissions are reviewed later.

- The fifth step is to prepare a first draft. This should be done quickly but completely. It is important to complete all sections even if the material is inadequate and the presentation is rough. The first drafts will highlight many of the major deficiencies in the document both in terms of content and presentation. There will in all probability be some major programmes of work needed to get the content right.

- The sixth step is to produce a second draft. This draft should contain few reservations about content and writing style. It should bear a close resemblance to the ultimate document, except that no executive summary will have been developed. At this stage proper diagrams, graphs, and photographs should be positioned correctly in the document.

- The seventh step is to put the document through a thorough management review. The first task is to have the document read from these standpoints:

- o by a management expert
- o by a technical expert
- o by a lay reader.

Their comments should be incorporated into composite changes which can satisfy all the readers. The second task is to challenge:

- the interpretation of customer needs
- the bid strategy
- the major themes in the draft proposal.

Adjustments should be made at this point, keeping the changes within the original proposal outline.

If this seventh step is carried out inadequately there will be no chance to retrieve the situation later. Moving to the next step essentially passes the point of no return.

- The eighth step is to produce a customer draft. This should simply be a clean well presented copy of the text which is structured in the way that the final document is planned. The purpose of this step is twofold:

- it allows the document to be assessed as such
- it provides an opportunity for a preliminary review with the client.

- The ninth step is to arrange for a review of the document with the customer or client. This review may be an unusual experience for the client and agreeing to it may meet with some resistance initially. However, it is important to make the point that the primary purpose of the review is to ensure that the submission is simply the most appropriate that can be assembled by the organization to meet the needs of the client. On occasions this step may be debarred by the terms of the invitation to bid.

  This client review will raise the issue of price. In general no prices should be reviewed or discussed. The purpose of the session is to check out objectives, scope and technical aspects of the proposed solutions.

- The tenth step is to amend the document draft as necessary following the client review, to write the all-important executive summary and prepare the final document. Whilst this document should reflect the needs of the client, which may involve constraints on size and chapter headings, it should also be unmistakably in the adopted house style of the organization.

Some general formats

A generalized format for a submission is given in Fig. 14. In practice

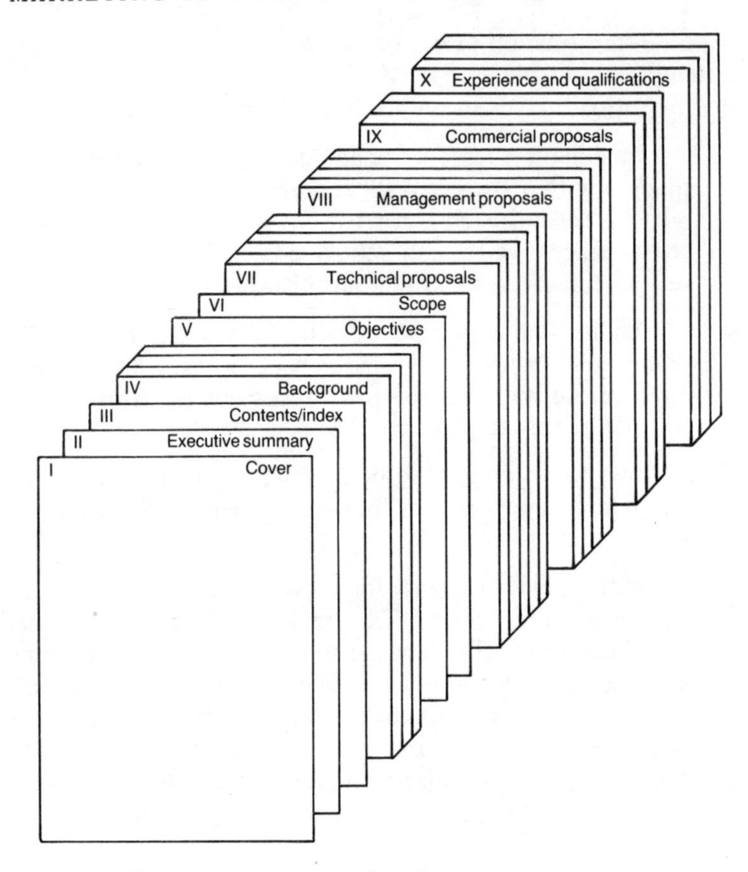

Fig. 14. Ideal format for a proposal

this would have to be adapted to meet specific requirements laid down by the client. The ideal document comprises ten sections.

The cover

- The cover of a proposal document is often a neglected area. In many instances it is merely an afterthought. At best it is often a standard organization cover designed to be anonymous. It should attempt to be:

- distinctive
- attractive

- clear in terms of subject matter
- identifiable with the prospective client and the bidding organization.

A generalized version which would form the basis of a suitable cover and be suitable to preserve a house style is given in Fig. 15.

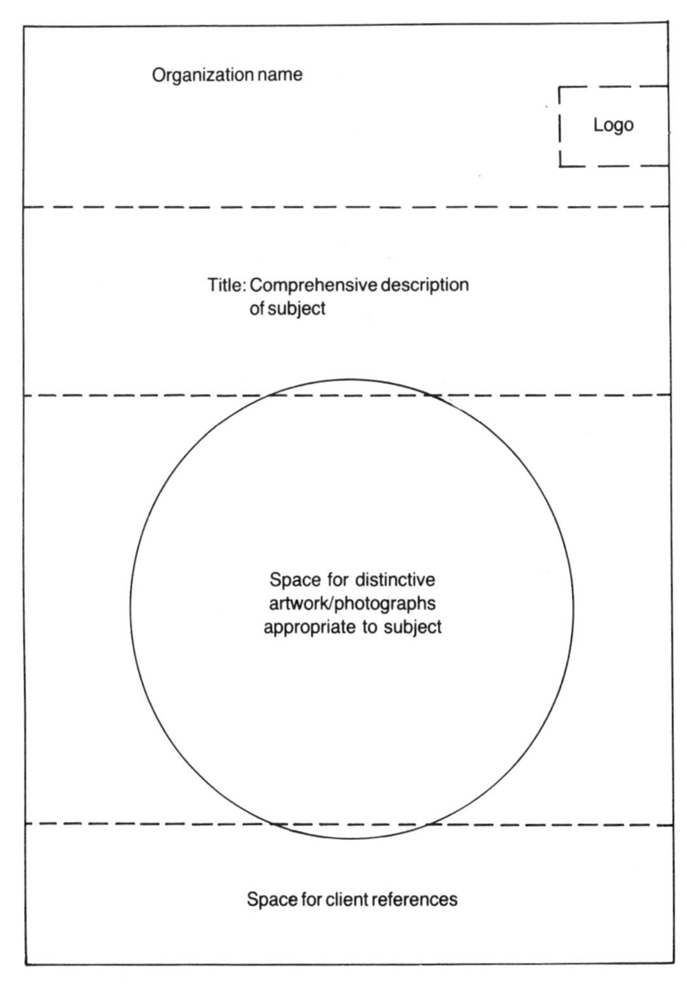

Fig. 15. Ideal format for proposal cover

The executive summary

This is the all-important step. The executive summary should be prepared only when the other sections are complete and after the client review stage. The executive summary should ideally be in the form of a letter of transmittal, that is addressed to one person and sent by one person. If at all possible it should be chief executive to chief executive.

- The summary should follow the rules of presentation in that the sequence of material should be:

  - outline of recommendations
  - key conclusions which have shaped the proposal
  - main reasons why the recommendations are felt to be appropriate, including relevant comments on price if a premium has to be justified.

The letter itself should ideally be about seven or eight pages long and certainly no more than fifteen. It should be written by the most senior person in the organization that will be involved and should be in his words. A good letter of transmittal serving as an executive summary will take two or three days to get right.

Occasionally, the terms of a particular submission call for an executive summary in the form of a separate self contained document. In this instance the same general rules apply although it will not be such a personal document and will be expected to be longer than a letter of transmittal.

Very few people on the client staff involved with the key decisions will want to flog through a detailed proposal. Instead they are likely to delegate its assessment to key technical staff whilst they probe various aspects of interest or concern. A well presented executive summary can make a tremendous impact with this group of decision makers if it communicates both good arguments and a checklist of key issues.

Contents/index

The contents or index pages should be a simple routemap through a document, referencing every headed section with appropriate page numbers. Some attention should be paid to making this a comprehensive listing so that easy referencing can be achieved. A document that is difficult to explore and study is very frustrating for the reader.

Background
This section is often missed out of proposal documents, and many clients will comment dismissively about its need. It has one very useful purpose. It is the only opportunity to play back to clients what has been understood about their organization and the background to their need. This section is valuable during client review sessions and often is the section that undergoes the greatest degree of editing and updating. When this section is well prepared it gives clients a strong feeling that their operations are understood properly.

Objectives and scope
These two sections should be short and fairly formal. Their purpose is to record the direction and context of the proposal submission. The objectives should be largely a replay of the client's briefing material whenever this is adequate. In drafting out a section on scope, it is important also to concentrate on those aspects which are specifically not to be considered. During the client review stage of the document, the section on scope also tends to generate a wealth of comments and alterations.

Technical proposals
- It may be that this section is heavily influenced by the instructions of the prospective client in terms of format. If this is so it usually reflects:

  - the client's view on priorities or important issues
  - the client's plans to evaluate submissions.

In general the technical proposals should start with a systems breakdown of the project or development under consideration, and then treat each element of the system separately. This treatment should start with clear technical recommendations, with supporting outline specifications, and be followed by a review of the reasons for the specific proposals.

This section has to be written initially by technical experts to be read and evaluated by technical experts. Subsequent editing should confine itself to issues of comprehension and readability.

Management proposals
- This section should concentrate on:

- the proposed management structure for the development concerned
- proposals for liaison and interim reporting
- programme and timetable, identifying major milestones for the client
- describing key support resources and their deployment.

The management proposals need to communicate competence and credibility. This section is also the only part of the proposal submission which can tackle any issues necessary to support a bid which is not expected to be least-cost. If the client is prepared to pay a premium for a superior solution, he has to find reasons in this section to justify this course of action.

Commercial proposals

It is beyond the scope of this book to comment in detail on commercial proposals. It is sufficient merely to state that this section should comprise details believed to be acceptable to both parties in formats which are also recognizable by both parties. If, however, there are major qualifications in relation to the strict terms of the invitation to submit proposals these should be clearly signalled and appropriate comments should be made.

Experience and qualifications

The preparation of this section also tends to leave a lot to be desired. Often it is merely a dump for all types of standard productions, sometimes produced on a wide variety of word processors in many different formats, and for company brochures and publications. This approach is not only untidy, it is also wasting an opportunity. This section should be developed specifically to support the submission and must be clearly seen to have been developed accordingly. It should focus on detailing experience under the following headings:

- *Team* – experience of individuals in the management team
- *Applications* – similar scope assignments previously completed
- *Technology* – uses of similar technology in all kinds of applications
- *Industry* – experience in working in the industry or service sectors concerned

- *Country* – relevant experience in working in the specific local working environment.

All details in this section should be well-marshalled and easy to read.

Speculative proposals

This chapter has essentially focused on preparing suitable documents to respond to invitations to bid in a specific situation. However, the main purpose of this book is to promote a marketing-led approach to business development. This approach will involve the early identification of opportunities and the prospect of acting before situations are opened up to conventional bidding. In these circumstances speculative proposals should be developed.

- Such a document should have two objectives:

  - to whet the appetite of the prospective client for a specific course of action
  - to structure from the very beginning the approach that ought to be adopted.

No client is likely to be interested in anything which does not meet a genuine need so the purpose of a speculative proposal is to demonstrate benefits to prospective clients, not to attempt to sell whatever services your organization has to offer.

A well structured document which is aimed at a genuine client need will establish the ground rules for any subsequent open bidding and could lead to a privileged position in attempting to meet these needs. No speculative proposal should, however, be a surprise to prospective clients. They should only be prepared when clients have indicated that they would be happy to receive a speculative document. Correspondingly, no speculative proposals should be developed if it is felt that the client will not be in a position to review the proposals and respond. In other words, speculative proposals should be more than interesting desk studies.

The format for speculative proposals should be based on the outline given in Fig. 14. More emphasis needs to be placed on the section on background and less on technical, management and commercial proposals when compared with other proposals. The section on previous experience should not be necessary at this stage and sections on objectives and scope cannot be written in formal terms. In some instances a speculative proposal may only require an executive summary section.

Submission document team

The key step in producing a first rate submission is to commit proper resources to the document preparation team. This needs to be done at the very beginning of the bidding process. It has already been suggested that the preparation of the document should be regarded as part of the research process. This means that the submission document team needs to be structured so that:

- the people likely to be involved in managing the work that would result from any successful proposals are involved in a way that reflects their relationships in a project management team;
- a specific document manager is appointed to assemble the contributions, edit them, and produce drafts in a common format. The document manager is not a junior team appointment, as the very process of preparing the material will give a unique insight into the submission and the bid strategy. If at all possible the document manager should be the anticipated project manager.

Summary and golden rules

When developing submissions, the golden rules are:

- have all materials prepared professionally;
- make all materials easy and enjoyable to read;
- organize an opportunity for a review with the client before final publication;
- aim to provide a document that is regarded as valuable by the client;
- be consistent in quality and appearance with other submissions prepared by the organization;
- do not divorce the document production from the process of working-up technical, management and commercial proposals;
- make the submission document the responsibility of the anticipated project leader.

10 People

People orientation

Marketing is an accepted element in professional and commercial management practice but less so in the civil engineering profession. In companies with a weak tradition in marketing, it is often their perception of what marketing people in other companies do that causes the greatest amount of confusion. It is important to recognize two types of marketing people, which for the sake of convenience will be known as type-A and type-B.

- *Type-A*

  This is the marketing specialist. He will see his whole career in marketing from the time he leaves college to the time he becomes chief executive. He is the prototype marketing whizz-kid. However, he will only be found in a small number of companies. These companies tend to exhibit characteristics such as:

  - fast moving consumer goods
  - easily identifiable purchasers and end-users
  - sales with a high proportion of impulse-buy decisions
  - requirements for fast feedback on sales achievement
  - marketing programmes capable of rapid change and redeployment.

These are not the characteristics one finds in civil engineering, or indeed in many other fields of business endeavour.

- *Type-B*

  This is the technical specialist. He will see marketing in strictly functional terms and looks on marketing as a career step activity. He will be found in many companies, particularly those exhibiting characteristics such as:

- capital goods/major projects
- no single purchaser to be found
- sales with a high proportion of multi-criteria purchase decisions
- very slow feedback on marketing initiatives through long lead times.

These characteristics are more common in civil engineering. In general, the need is not for marketing specialists, but for technical specialists with a high degree of marketing competence. The technical specialists are ultimately the only people who can earn credibility with clients, although they will need the advice and assistance of professional marketing staff in areas such as PR and graphic design under the direction of specialists.

The marketing mission in civil engineering is to be able to work with type-B people and to turn them into effective marketing resources.

Essential prerequisites for success

- The ability to turn type-B people into effective marketing resources requires attention to:

  - attitudes to marketing as an activity
  - the ability to measure performance credibly and in a way acceptable to the marketing resources of the organization
  - professional training in marketing techniques.

Attitudes to marketing

- In civil engineering circles, marketing tends to suffer from the following attitudes.

  - Marketing is a diversion from the real task of doing work.
  - In a perfect world there would be no need for marketing as customers and clients should know who is best placed to serve them.
  - Marketing is a repository for failed project managers.
  - Marketing is a qualitative rather than a quantitative activity and defies rational measurement of performance.

In reality many of these attitudes probably spring from a feeling of discomfort about marketing by professional engineers and it is this aspect that has to be tackled in changing attitudes. It is essential for all technical specialists to see marketing as a career step in the sense that no-one can make it to the top of an organization without some experience in marketing. After all, the chief executive is the single most important marketing resource in any organization and needs to be prepared for this role accordingly. If this career step approach is recognized and emphasized there will be no shortage of ambitious engineers looking for suitable opportunities to develop their marketing skills.

Whilst recognizing the career-step approach to marketing, it is important not to forget that the marketing activity itself has to be organized as a permanent and ongoing process within an organization and cannot be left at any time without proper resources.

Performance measurement

There is an old adage of industrial engineers which says 'if you cannot measure it, you cannot control it'. Measurement in marketing should be carried out at three levels:

- organization level
- objectives level
- budgets level.

Organization level

- Simply stated, measurements at this level should establish whether plans laid down were subsequently carried out for:

  - the development of the marketing organization
  - the programme of marketing priorities.

No organization should be allowed to evolve with only few attempts made at corrective action, and least of all a marketing organization. Plans for recruitment, deployment and training should be measured against the outturn and conclusions need to be drawn from this experience about subsequent development plans.

The deployment of resources should be measured specifically against the programme of marketing campaigns and priorities.

Imbalances here should lead to questions either about resources or about priorities. These imbalances may never be corrected but they must never go unchallenged.

The measurement of performance at an organization level should result in a narrative document produced annually which comments on organizational development and resources deployment to meet programmes and objectives.

Objectives level

Marketing objectives will reflect the overall objectives of the organization in terms of financial performance and growth. Marketing has the task of positioning the organization so that these broad business objectives can be met from a specific set of clients, products, services and facilities. Performance measurement at the objectives level involves the review of:

- current marketing strategies
- current markets status.

- Marketing strategies will be developed later in the sections of this book that relate to different aspects of civil engineering. However, all marketing strategies have common features. In general, they comprise statements and quantifications of:

- market segmentation for the organization
- market penetration performances
- product/market developments and achievements
- diversification activities.

The best approach to measuring these strategies is to use an approach known as SCAD analysis. SCAD is the mnemonic for segment, concentrate and develop. This is the process for creating marketing strategies. Each market should be segmented into product/service categories and for each segment a specific plan should be developed. The purpose of segmentation is to focus on a purchaser group or market that is relatively homogeneous in its purchasing behaviour so that common marketing strategies would be appropriate for different purchasers in the segment.

The SCAD approach should provide the basis to review market segmentation and associated market development plans. Once a market segment definition has been adopted a measurement of market penetration or market share should be made for the

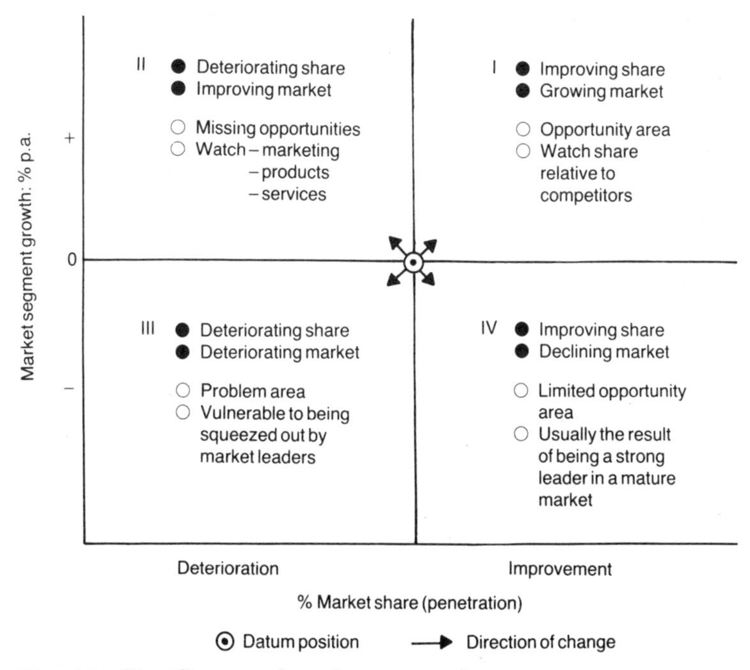

Fig. 16. Significance of market penetration

organization in each market segment. Market penetration is a key management parameter and also needs to be established for competitors for complete understanding. The relative movements of market penetration and overall market growth provide important signals about past marketing success and future prospects.

Figure 16 shows the interrelationships and the implications when market penetration changes. This review is valid for both existing and prospective diversified activities. Diversifications, however, are only justifiable in region I of Fig. 16 if organic growth is planned. If growth by acquisition is planned it is possible to work in regions I and III of Fig. 16.

- Current markets status can be established using an approach known as SWOT analysis. SWOT is the mnemonic for strengths, weaknesses, opportunities and threats. This analysis is remarkably simple and remarkably powerful. It needs to be applied to the following aspects of the business:

- o products/services
- o pricing/tendering
- o distribution/mobilization
- o promotional activities
- o marketing/selling resources.

A separate SWOT analysis should be carried out for each product or service, whilst a combined analysis is appropriate for each of the other activities. This analysis is very good at identifying priorities for support and priorities for attention.

Budget level

- The marketing budget should comprise details on:
- o sales targets
- o sales conversion ratios
- o submissions targets
- o staff deployment and costs
- o other management expenses.

One aspect of performance measurement is to compare out-turn with budget plan, and to relate differences to changes in workload assumptions and to success factors in securing work or meeting objectives. This process will help in budgeting for the future and in establishing standards of performance for marketing.

When marketing efforts have stabilized and are meeting the needs of the organization there will be a steady state effect in terms of:

- the backlog of work oustanding needed to meet the volume needs of the organization;
- the sales conversion ratios required to provide adequate revenues;
- the utilization of marketing resources on prospects selected for development.

These parameters are different for different organizations and in some ways characterize the essential differences between organizations.

Training

The key to successful marketing in a civil engineering environment as with other disciplines is probably training. All staff should market an organization. They should be trained to do it well and should always feel comfortable in doing it. The training itself should have two aspects:

- formal training
- experience training.

Formal training

- Formal training programmes need to address:

  - principles of marketing
  - structural approach to marketing
  - making presentations
  - communications through documents.

These items are very much the ones with which this book is primarily concerned. However, all material aimed at self-development is limited and becomes much more powerful when supplemented by attendance on formal training courses.

Identifying suitable training courses can be a major problem. Often the full relevance of a course is in doubt as many courses are designed to appeal to a wide range of business groups. For those courses aimed at senior management this should not be a problem. People attending should be able to establish relevance and may well benefit from being exposed to other course members from different business backgrounds.

The problems are much more severe when junior staff are involved, and there is no better solution than to have a group of professionals work with the organization to design and run a series of specific training programmes relevant to the organization.

The principles of marketing and the structural approach can be taught and illustrated. Being competent in presentations and documentary communications involves the organization of practice and there is no substitute for this.

Experience training

All recently trained staff, in whatever subject, will look for immediate opportunities to use their new-found skills. If people are not used immediately after training in a formal marketing role, then opportunities need to be created for them to play a part in occasional marketing activities. Typically, the best opportunities to do this will come from their participation in two activities which are developed later:

- specific marketing campaigns
- marketing audits.

11 Segmentation

Putting marketing into practice

- The practicalities of marketing, in common with other management tasks, require:

  - setting objectives
  - creating a programme to meet the objectives
  - organizing resources for results
  - measurement of achievement
  - review of success and modification of objectives, programmes, or organization as necessary.

This generalized approach differs significantly in practical terms from one consultant to another, from one contractor to another and from one public servant to another, but it is useful to consider each of these groups as if they were broadly homogeneous.

However, before specific approaches can be reviewed a number of purely marketing issues have to be appreciated. The most fundamental of these is the concept of segmentation.

Concept of segmentation

The successful approach to marketing is one of:

- singling out key purchaser groups for attention;
- subdividing these groups into segments that broadly behave in the same way;
- getting to know these segments in great detail particularly with regard to:

  - what current needs are
  - what are regarded as key issues for the future
  - how purchase decisions are influenced;

- Choosing for attention, those segments which:

  - offer growth prospects
  - could be served from existing resources and skills;

- Developing products, services and distribution channels to make the purchase decisions easy in these chosen segments.

Segmentation, it can be seen, is the process of cutting up the business into manageable parts where each part has a strong customer-group focus.

This approach sounds perfectly logical but effective marketing is still beyond the reach of many people and organizations. When this is so the problem tends to be one of understanding and applying the concept of segmentation. Unless organizations and businesses can segment their markets, so that specific actions can be concentrated there, much of the marketing effort will be wasted.

Dimensions and scope of segmentation

Segmentation can be regarded crudely as a matrix linking services or products offered to different customer, client or purchaser groupings. However, this is not quite so simple in practice. The problem comes in exercising judgment on how to divide up the service offerings and the client groupings so that they reflect key features of the market.

- An outline market segmentation for consulting engineers is put forward in Fig. 17. It is a first cut only and will not reflect the important subtleties of:

  - earnings potential
  - geographical or regional differences
  - competitive pressures
  - volume growth expectations
  - particular resources or strengths of an organization.

Each cell in such a matrix will have to be broken down further. If the local authorities/engineering design cell is developed so that specific service offerings are identified, the result would be something like that shown in Fig. 18. It is in this step that the distinctive detail that is characteristic of every organization comes through and allows unique segmentation to be developed. Typically, segmentations are developed for different regional markets.

Client groupings	Service offerings				
	Engineering design	Preparation of bids	Resident engineer	Feasibility studies	Multi-client reports
Public sector ○ Government agencies ○ Local authorities ○ Specialist authorities					
Private sector ○ Mining and resources ○ Energy and utilities ○ Manufacturing ○ Retail and distribution ○ Transportation					
Consortia ○ Design and construct ○ Investment					

Fig. 17. Outline market segmentation

It is good practice initially to mark up the detailed cell segmentations in one of these ways. To show areas of interest and focus (marked with *x*), to show areas that will be clearly declined (marked with *o*), and areas with no presence or experience which could be of interest (left blank).

Segmentation as an analytic tool

When the segmentation exercise is completed it is possible to review strategic and marketing objectives. The segmentation exercise is a fundamental and highly significant task for all businesses. If it is being done for the first time, it will produce many surprises, particularly about relative strengths and weaknesses. It is a process not to be rushed nor skimped and could easily cover a table top in terms of physical size.

Local authorities	Engineering design				
	Roads	Bridges	Services	Tunnels	Etc.
○ Highway Department	x	x		x	
○ Public Health Department			x	x	
○ Housing Department			x		
○ Parks Department					
○ Transportation Department		x	o	x	
○ Etc.					

Fig. 18. Example of detailed segmentation

- The first step in the analyses is to establish some key business data for each of the detailed segment cells. This data should cover:

o market size and growth
o profile of own revenues and those of key competitors
o profile of own performance and those of key competitors.

An example of this data development is given in Fig. 19 for the Highways Department/Road Engineering Design cell. It gives a basis for analysis.

For each of the cells a SWOT analysis should be carried out to establish its importance as an area for business development. If business development efforts can be justified then each of these cells will demand a subtly different approach and this becomes the basis of the marketing plan for an organization. An example of this type of analysis is given in Fig. 20 again for road design in the Highways Department.

- The whole purpose of segmentation is to produce a structured

Segment: *Highways Dept: Roads engineering design*

Key data

1 Market size and growth (UK)

	Datum	Y+1	Y+2	Y+3	Y+4
National programme	£300 m	£250 m	£260 m	£300 m	£260 m
Design content	£15 m	£14 m	£13 m	£13 m	£13 m
Prospective clients	165	165	165	130	130

2 Revenues profile

	Y−2	Y−1	Datum	Share
Own revenues	£1 m	£1·5 m	£1·5 m	10%
Competitor A	£3 m	£3 m	£3 m	20%
B	£1 m	£2 m	£1·5	10%

3 Performance profiles

	Own	A	B
Gross margin	25%	30% (E)	20% (E)
Return on funds employed	20%	22% (E)	22% (E)

4 Resources deployment

Assets – fixed	£1 m	
– net current	£2·0 m	
People	11	
Skills	Engineers	7
	Support	4
Deployment	Marketing	2
	Design	9

Fig. 19. Typical form of segment data development

framework for business analysis and to aid the process of setting strategic and marketing objectives. The segmentation process provides the physical framework and is a convenient device to analyse:

- market status
- competitive position
- SWOT implications
- resources deployment
- performance results and expectations.

In this way a consistently structured picture of the market and the competitive position of the organization is developed for each

Segment: *Highways Dept: Road engineering design*

Strengths	*Weaknesses*
○ Strong position with local authorities in general ○ Specialization in fly-over and viaduct designs	○ Poor position in Scotland ○ Not the market leader
Opportunities	*Threats*
○ More urban motorway work requiring elevated decisions ○ Greater emphasis on urban road improvements	○ Downturn in public spending programmes ○ Over-capacity in the sector

Implications
○ Target – Urban authorities – Elevated sections/interchanges ○ Reduce dependence on sector for income

Fig. 20. Example of segment analysis

of the cells which in essence are the fundamental building blocks of the organization. Segmentation is one of the few tools that is unique to marketing and is all the more powerful for that. Marketing tends to be a process of orchestration rather than something with its own science and consequently the general approach tends to be all-important. The acid test on segmentation is that when it is carried out correctly, the segment names will become the common business language of the organization and the segmentation will appear to be quite natural or obvious.

In most organizations, a number of segments may share common resources and facilities. When this happens they tend to form natural organization or management units and are often referred to as Strategic Business Units (SBUs). The significance of the SBU is that it should be a profit centre in organizational terms and plans and budgets should be developed at this level over the full planning horizon of the organization.

12 Performance analysis

Measurement of performance

- The ability to measure performance is critical in:

  - identifying areas of business focus
  - setting objectives and targets
  - projecting future results
  - controlling resources and activities.

Sometimes, it appears that measures of performance and results can be quite different for different aspects of operations. Whilst there may be superficial differences and whilst different parameters may be used in different circumstances, it is essential that all measures of performance are linked numerically to the overall objectives of the organization. The logical place, therefore, to start with performance measurement is with the performance objectives of the corporate body.

Corporate measures of performance

- There are two distinct cases to consider here:

  - commercial organizations
  - public sector organizations.

Commercial organizations simply have to provide attractive returns to their owners or shareholders whilst meeting market needs. For public sector organizations it is not usually so simple. Where an organization in the public sector has been structured on a trading fund basis, such as the Royal Mint, a straightforward commercial approach to performance measurement can be adopted. Where an organization has to provide a public service, such as the Offshore Supplies Office of the Department of Energy, the commercial approach to performance measurement is much less relevant.

However, in the UK in recent years, the movement of organizations in the public sector into the private sector has reinforced the practice of applying simple commercial measures of performance.

External and internal perspectives

The shareholder, who has an external perspective on the performance of an organization, is typically concerned with earnings per share (eps). Management, who have an internal perspective on performance, are typically concerned about return on funds employed (ROFE) or return on total assets (ROTA). These measures are linked as shown in the mathematical note in Fig. 21.

Line	Item	Abbreviation	Relationships
From profit & loss account			
1	Sales	Rev	
2	Cost of sales	COS	
3	Management costs	Man	
4	Operating profits	PBTI	= (Rev)–(COS)–(Man)
5	Interest charges	Int	
6	Tax	Tax	
7	Minorities interests	Min	
8	Shareholders profits	PATI	= (PBTI)–(Int)–(Tax)–(Min)
From balance-sheet			
A	Shareholders' funds	Sha	
B	Borrowings	Bor	
C	Total assets	TA	= (Sha)+(Bor)
D	Number of shares issued	nsi	
Measures			
I	Operating margin	opm	= (PBTI):(Rev)
II	Net margin	mar	= (PATI):(Rev)
III	Asset turn	AT	= (Rev):(TA)
IV	Return on total assets	ROTA	= (PATI):(TA)
V	Return of shareholders funds	ROSF	= (PATI):(Sha)
VI	Gearing		= (Bor):(TA)
VII	Earnings per share	eps	= (PATI):(nsi)
Other relationships			
X		Int	= f(Bor)
Y		eps	= f'(ROSF)

Fig. 21. Corporate performance measures

Factor	Examples: % p.a.			
	High risk	Average risk	Low risk	
Interest rate	10	10	10	
Business premium	15	10	5	
○ target ROTA	25	20	15	
Inflation	(5)	(5)	(5)	
○ real ROTA	20	15	10	*Internal perspective*
Gearing (@ 70%)				
○ Return on shareholders' funds	28·5	21·5	14·5	*External perspective*

Fig. 22. Developing business performance objectives

The main difference between the external and internal perspective concerns the role of borrowings and consequential interest charges. The level of borrowings in relation to total assets is known as gearing and the subject is more properly part of corporate finance. However, it is sufficient to say that decisions taken on borrowings affect the performance measures for an organization and this has implications for marketing objectives.

Using the conventions in Fig. 21

$$\text{ROTA} = \frac{(\text{PATI})}{(\text{TA})}$$

$$= \frac{(\text{PATI})}{(\text{Rev})} \times \frac{(\text{Rev})}{(\text{TA})}$$

$$= (\text{Mar}) \times (\text{AT})$$

i.e. (Net margin) × (Asset turn)

This corresponds mathematically to

$K = (x) \times (y)$

When K is kept constant this relationship is an hyperbola, and for different values of K a family of hyperbolae can be developed. This process is shown in Fig. 24.

Fig. 23. ROTA mathematics

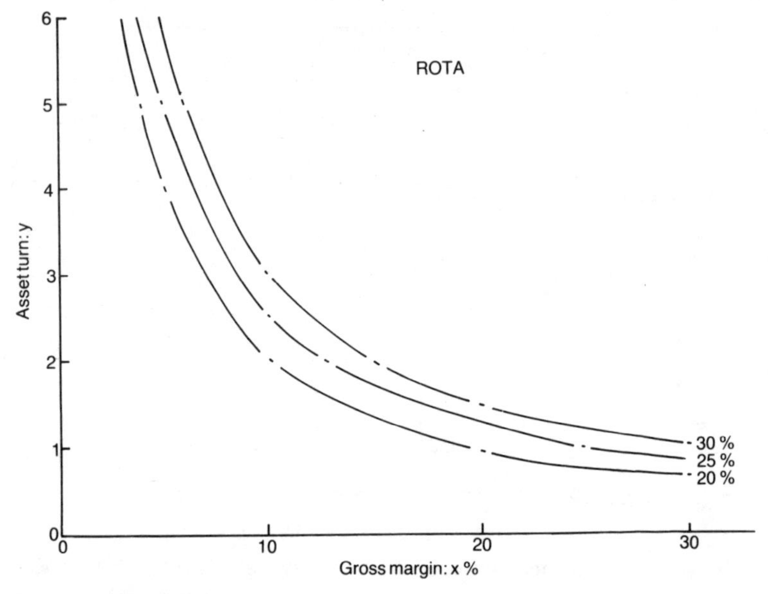

Fig. 24. ROTA curves

Setting corporate performance targets

The ROTA objectives of an organization will be related to prevailing rates of interest, to prevailing inflation, and to a subjective premium for risk. An example of this mechanism in setting objectives and the implications for return on shareholders funds is given in Fig. 22.

The behaviour of ROTA objectives is worthy of some further analysis. It is a measure that truly combines the results of the profit and loss account with the items on the balance sheet. It is a two-dimensional measure, and there are an infinite number of ways in which an organization can meet its ROTA objectives in an equally valid way. In practice, organizations are constrained by the nature of their operations and their immediate past performance. The mathematics of ROTA analysis are given in Fig. 23 and its graphical representation is given in Fig. 24.

Use of ROTA curves

- Every business has its own unique mix of margin and asset turn to produce a given ROTA. Broadly speaking:

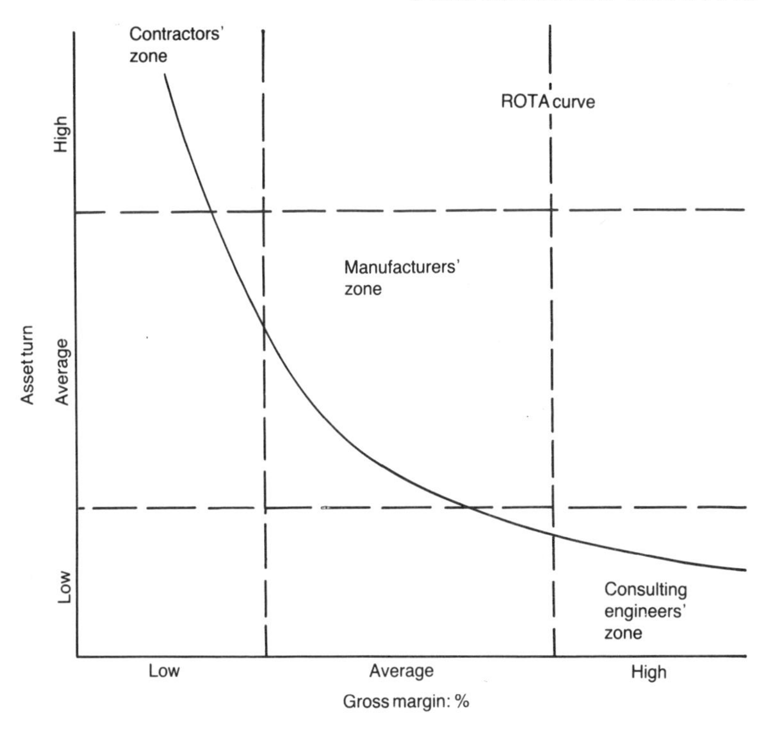

Fig. 25. Performance classification

- contractors have low margins and high asset turns;
- manufacturers have average margins and average asset turns;
- consulting engineers have high margins and low asset turns.

This separation is shown clearly in Fig. 25.

Immediate past performance by activity type should be plotted on a ROTA curve and measured against the target ROTA line. If, for example, there is a shortfall in performance then corrective action is required. The best course to adopt is called 'the path of least regret' and an example of this is given in Fig. 26. It is the shortest line from current performance to the ROTA curve at the point where it makes a right angle with the tangent. Corrective action can now be quantified in terms of an asset turn improvement and a margin improvement.

Action programmes can now be detailed through a technique known as the *Improvement Tree*. A start should be made with the

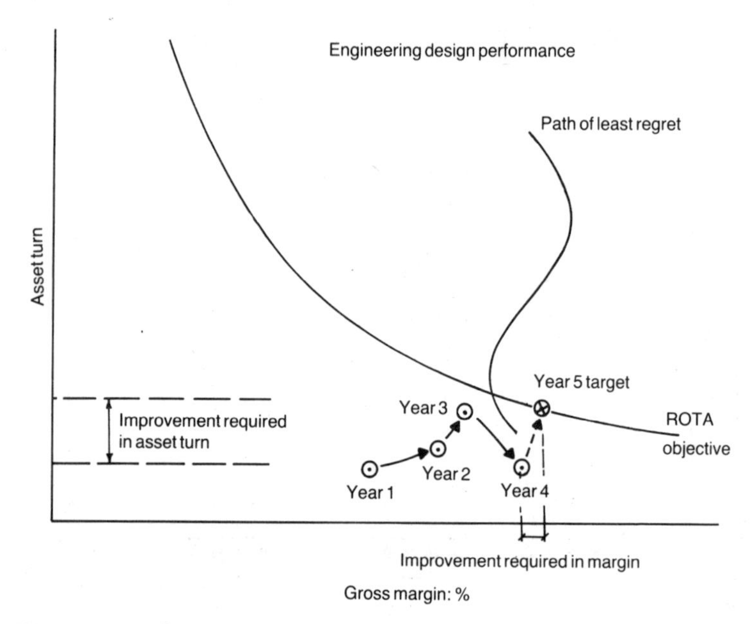

Fig. 26. Performance objectives

assets side of the business, as these are less people-intensive and usually more under direct management control. Then the margins side of the business can be tackled. The value of the improvement tree lies in its structure. All activities, if carried out, will improve ROTA performance and will not be affected by other supporting actions. A simplified example of the tree is shown in Fig. 27.

The ROTA curves can also be used to track the performance of competitors. However, there are usually some practical difficulties in doing this if a competitor is a broadly-based organization with activities in a number of very different markets. When the ROTA analysis is tied into the broad segmentation of the business and the competition is analysed in the same detail, a powerful tool emerges. The biggest practical barrier to the analysis of competitors is usually the lack of information about the deployment of funds and assets behind each part of the business.

Strategic development

The ROTA analysis is most useful in setting targets and in analysing immediate past performance with the objective of initiating short

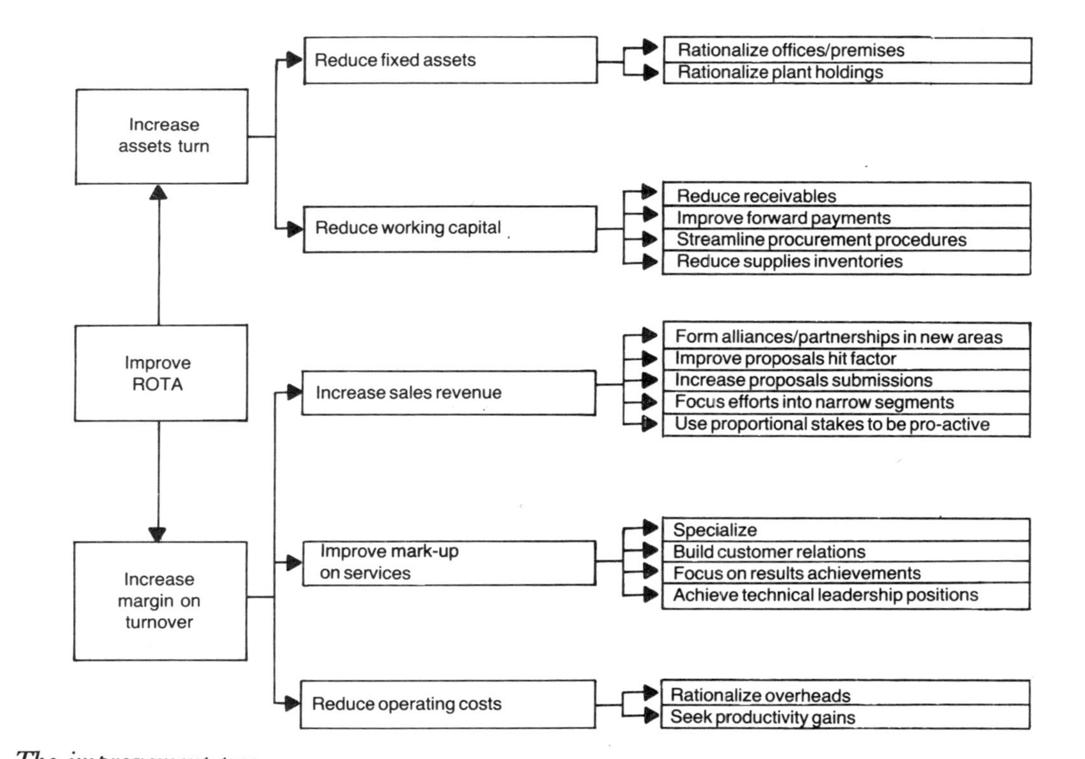

Fig. 27. The improvement tree

term corrective action. A broader measure is needed to address longer term issues and the whole subject of strategic development. However, before looking at how these situations can be analysed and assessed it is necessary to develop some background to current thinking in the field of strategic studies.

There are two key principles behind modern business strategy analysis:

- that privileged commercial positions come from leadership of some form;
- that leadership positions come from either:
  - the exploitation of some facet of activity which gives an overwhelming competitive advantage, or
  - the advantages of experience or cumulative production on unit cost reductions which give price leadership.

It is beyond the scope of the book to discuss these concepts in detail, but it can be seen that if these principles are adopted then strategic development is the process of identifying some aspects of the marketplace where existing resources and experience will give the organization some critical competitive edge.

Portfolio analysis

The argument that leadership positions ultimately provide cost advantages and consequently price leadership has focused attention on the importance of market share. The argument continues that if experience leads to greater commercial advantage then those organizations with the greatest accumulation of experience will have the greatest advantage. A measure of this effect is market share, and so the argument concludes that those organizations with the largest market share in a business sector should have the greatest opportunity for price leadership and profitability. This message was delivered in no uncertain terms to the major multi-national manufacturing corporations during the 1970s and led to a scramble for market share.

The classic form of analysis in this area is the portfolio analysis developed and promoted by the Boston Consulting Group and often known as the Boston portfolio matrix. An example is given in Fig. 28 of an industrial contracting conglomerate. In the conventions used, the size of each circle, which corresponds to a business

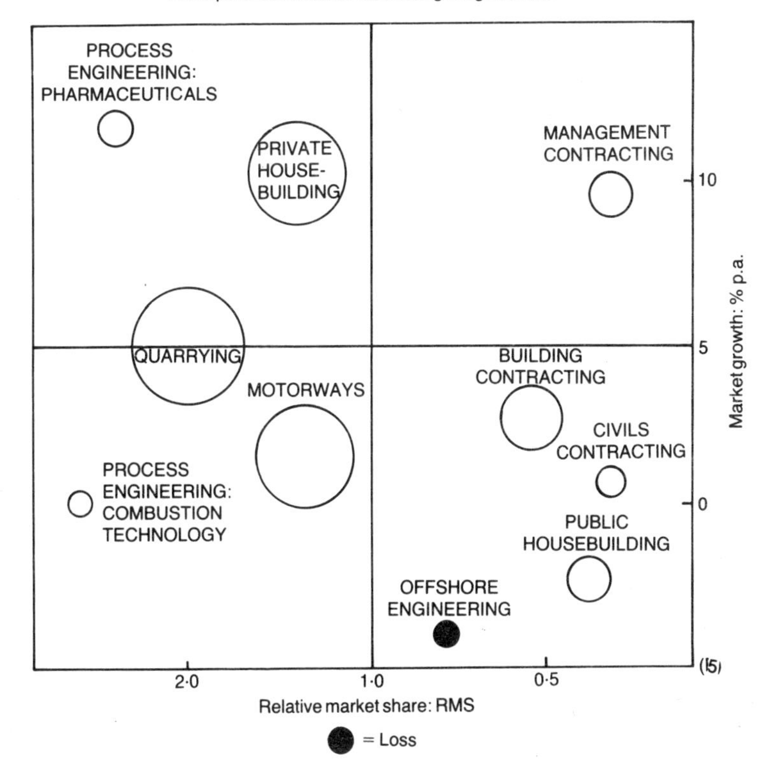

Fig. 28. Portfolio analysis

segment, represents the profits. In this example the business analysis would be as follows:

- *Quarrying:*
  - Market leader (RMS is greater than 1.0)
  - Steady growth (less than 5% pa)
  - Massive profits stream
  - *Major cornerstone of the business*
- *Private housing:*
  - Market leader but others possibly of similar size
  - Good growth

- o Good level of profitability
- o *Major profits generator*

- • *Process engineering – pharmaceuticals:*

- o Market leader by a significant margin
- o Very high market growth
- o Modest level of profitability
- o *Major development opportunity*

- • *Process engineering – combustion technology:*

- o Market leader by a significant margin
- o Market in decline
- o Poor profits
- o *Questionable future unless leadership position can be made to count*

- • *Motorways:*

- o Market leader although others of similar size might be present
- o Market in decline
- o Reasonable profits
- o *Leadership being used to protect profits in weak market*

- • *Management contracting:*

- o Not the market leader and less than half the size of the leader
- o Good growth
- o Reasonable profits
- o *Key decision is whether to support the activity to win greater market share: prospects good*

- • *Building contracting:*

- o One of the leaders in the field
- o Steady growth
- o Reasonable profits
- o *Business worthy of support*

- • *Civils contracting & public housing:*

- o Relatively small participant
- o Poor market growth prospects

- Only adequate profits
- *Should consider withdrawal if resources can be used to support other activities*

• *Offshore engineering:*

- One of the leaders in the field
- Serious market decline
- Unacceptable losses
- *Major candidate for rationalization.*

This sort of analysis, which is one of an increasingly wide range being adopted in business, leads directly to a plan for strategic development and it is this plan that usually sets the framework for the marketing task in an organization.

13 The consulting engineer

Role

The generally accepted role of consulting engineers comprises:

- to play a part in project identification for clients through feasibility studies, economic analysis and planning studies;
- to provide specialist design services to clients;
- to prepare project plans for tendering;
- to provide supervision of contractors through the practice of resident engineers;
- to provide key professional services to the industry on technical, contractual and planning matters.

This role has evolved over many years and fits in with the practice of clients securing the services of consulting engineers to design their work and represent their interests in a professional manner.

Evolution of the profession

- The whole practice of civil engineering in the UK has evolved in a way which:
  - allowed a proper division of professional risk between the different work packages that make up a project;
  - allowed a fair and competitive price to be determined for a project in the tendering process;
  - allowed a fair adjudication of the contract between the client and the contractor.
- The advantages of this practice have been manifold.

- o It has created specialist consulting engineering practices in design terms – this, in turn, had led to a number of British consulting engineers being regarded as world leaders in their field.
- o It has ensured that clients secured projects at minimum cost and minimum risk in completion terms.
- o It has provided a proper mechanism for the unforeseen to be properly accounted for, through the practice of variations payments.

However, this traditional role is now coming under increasing pressure. Clients, particularly those in the private sector, are most concerned about overall cost and are looking to a greater extent for all-embracing fixed-cost bids for projects. There is a growing preference for design and construct bids. The move in this direction has been accelerated by the cynical exploitation of the traditional system by some contractors who adopt the strategy of putting in low unrealistic bids to win work and then exploit the variations to the full to make up an element of profit. Private sector clients who may, only once or twice in a working lifetime, embark on a major project can be horrified at escalating costs which cannot be staunched by their consulting engineers.

Future trends

These changes have by no means run their full course yet. In future the consulting engineer will have to work in a number of different ways:

- o a traditional professional role
- o as part of a bidding consortium
- o as a professional on the staff of a contractor.

- The traditional role will tend to focus on:

- o increasing levels of design and services specialization
- o increasing dependence on public authorities who place a greater value on proper public accountability than total overall cost;

- On design and construct work, consulting engineers are faced with entering into consortia with contractors. This has a number of fundamental consequences:

- o bids will have to be supported by full designs if cost estimates and risk assessments are to be accurate
- o the design work will only be paid for through the successful bid;

- The design engineer inside the contractor will experience the problems of handling a wide variety of work and of being unable to settle into a specialist area to develop his skills progressively, at least to the extent possible in an independent consulting engineering practice.

One issue which is accompanying these changes is the competitive pressure on fees structure. This is developing into a growing concern in the profession about sources of fee income and about marketing practices. What may well result is a new wider professional role met by larger practices that are:

- o more flexible in business relationship terms
- o multi-disciplinary in skills and more specialist in services.

The key results areas

- The professional consulting practice should concern itself about three operational parameters:

- o the extent of proposals or bids outstanding, measured in value and in the number of weeks of potential work at standard weekly outputs;
- o the conversion ratio on bids, measured as a percentage of the value of work won of the value of proposals submitted;
- o the revenue realization of professional staff, measured as a percentage of realizable income at full output.

Marketing relates to the opportunities to submit bids and proposals. Selling relates to conversion ratios, although many people erroneously regard this as part of marketing. Profits are a consequence of staff realization and these are maximized in the long term when people are used effectively to secure client satisfaction. In this latter sense, performance becomes a weapon to use in marketing.

Each consulting practice should develop its own operational parameters and measure performance against each. If it is not currently done in a practice, it is a salutory exercise to measure the three parameters discussed, and consider their implications.

Key marketing themes

The primary purpose in measuring these key performance parameters is to quantify the size of the marketing and sales tasks for a company. This process was outlined earlier in Chapter 5.

To become more effective in marketing, a consulting engineering practice will need to undertake a programme with the following features.

- Prepare comprehensive listings of potential clients in those areas where services are to be offered. These listings should be categorized by service type and client type, and should be properly cross-referenced. A potential client is a person or group that can purchase services.
- Prepare comprehensive listings of decision influencers under the headings of:

  - government agencies
  - political lobbies
  - in-client groupings
  - research organizations or centres of excellence.

  These listings again need to be cross-referenced with service offerings.
- Research past activities of potential clients so that knowledge is gained about:

  - business or operational objectives
  - spending patterns
  - results and past achievements
  - preferences for partners and suppliers.
- Having undertaken the steps above which are preparatory, the next step is to create a programme to meet people from the first two groups. The objectives in meeting these people should be to:

  - get better acquainted
  - learn more about their activities
  - inform them about your activities and successes

- create a relationship which will allow them and encourage them to contact you in the future.

Provided proper preparation has been carried out this is the critical marketing step. Unless a specific issue is raised, never adopt a selling posture nor seek a commitment of any kind during a marketing activity. The idea is to become more knowledgeable about prospective clients and to establish credibility with them.

Follow-up visits, provided they are not too frequent, will enhance this position further. This process is fundamental in positioning a client so that he feels confident and comfortable to use your services.

- Support the field initiatives with programmes which reinforce the marketing message of the practice. Typically, this will involve activities such as:

- producing brochures
- advertising
- participating in exhibitions
- writing learned papers
- participating in conferences.

Many people see these activities as being the central features of marketing and expect better marketing results when more money can be spent on these items. This viewpoint is again erroneous, and large sums of money can be wasted unless a well thought through plan is developed. It is important never to forget that these activities are there to support the primary activity which is to meet influential people and make an impact personally.

The best way to approach these support activities is:

- to justify a budget for each, recognizing that exhibitions and advertising should be allocated whatever budgets are left after the other items have been covered;
- never to compromise standards in any way as, in professional life, only the best can be justified.

A good way to justify a budget is to relate it, as a policy, to turnover and then ensure that maximum value for money is realized.

- Invest in a relevant R&D programme to keep your technical leadership positions protected. This should not take the form of primary research, unless the practice is a research association which sells contract research. It should be a programme to maintain awareness of new technical developments worldwide through library research and participation in conferences. New ideas should be assessed for relevance, turned into practical perspectives whenever possible, and then promoted positively in practice communications and presentations.
- Invest in first rate document production capabilities and use a clear, distinctive and attractive corporate style and use it consistently.

Essential staff work

A successful marketing programme implicitly requires that not only are marketing tasks carried out professionally, but that the management effort and the resources of the organization are directed towards the right objective. Segmentation and business analysis are essential staff activities to ensure this happens.

14 The contractor

Role

The generally accepted role of the contractor comprises:

- to offer a service, typically to construct a facility to a specification, for a fixed price on a given bill of quantities subject to agreed exceptions, and to accept the risk for completing the work to time, specification, and profit expectations;
- to act for a client in helping them to secure the physical realization of their business plans.

Contractor companies have evolved in ways which have reflected their technical experience and expertise, plant investments and inventories and geographic presence.

Evolution of the business

Many contractor companies have grown from private entrepreneurial businesses into major multi-national organizations over the past 100 to 150 years.

- The underlying pattern to this development in the UK typically has been along the following lines:
  - a start in housebuilding or property development;
  - a gradual move into civil engineering work;
  - a gradual move into world markets starting with the Dominions and colonial possessions.
- The whole business area of contracting has broadly developed a common attitude in the UK towards its preferred methods of working. The generally accepted position is:

- never invest in a project merely to secure work;
- specialize in assessing risk;
- manage cash flows very tightly;
- make special contractual provisions for all areas of uncertainty and for all areas where the guidance of professionals is mandated.

Again this approach has fitted in well with the role of the consulting engineer and it has resulted in many decades of successful work and satisfied clients.

The critical impact of the public sector

Civil engineering is virtually synonymous with the construction of infrastructure. Whilst in the last century the monuments to infrastructural development in the UK, USA, Canada, India, South Africa, Australia and Argentina were largely privately inspired (often with British capital) by the late 1930s these developments worldwide had become the province of the public sector.

The British approach to civil engineering, because of its pragmatic approach to the division of risk between consultant and contractor, and because of its basically honest business approach, was a very attractive model for public authorities worried about their public accountability.

In contrast, civil engineering work from the private sector became more associated with process-related developments where design risks were fundamentally different in nature. Many American contractors grew out of their process design origins, and in the private sector, the American approach has predominated, particularly with the process-based multi-nationals.

Future trends

The future will be determined by the outcome of three trends evident now:

- the reduction in public expenditure worldwide and the search for private solutions to infrastructure needs again;
- the growing attraction of the American design and construct approach to many prospective clients;
- the diversification plans of UK contractors which are differentiating many of the companies that were formerly broadly the

same. The major UK civil contractors have, separately, become specialized in such diverse activities as:

- aggregate production
- mining
- shipbuilding
- fabrication engineering
- property development
- hotels and leisure.

These developments are being forced by the breakdown of the international market, particularly in its traditional form, and companies are seeking new business operations. This will give each major civils contractor a distinctive flavour to his marketing stance.

The key results areas

If the pivot of the contractors business is 'risk assessment' then the corresponding perspective of the client must be 'insurance'. Clients will look for:

- comfort, through credibility and competence
- least cost, for accountability
- easy personal relationships.

Public sector clients will place great emphasis on least cost and competence, whilst private sector clients will place a corresponding emphasis on achieving target product unit cost reductions with the greatest degree of certainty.

- The key results areas in the contractors business are:

- getting access to new work opportunities at an early enough stage to form a distinctive approach
- being able to influence the tendering procedures so that only a manageable number of tender invitations are issued
- emphasizing the 'insurance' aspects of the bid and building client confidence.

Key marketing themes

Success in marketing as a contractor will come from an ability to project a unique selling proposition (USP) to the market in general. The identification of a USP and its acceptance within a company is an important mission to undertake. However, it is a double edged weapon as not only does it tell the world what a company is good at, it also, by implication, tells the world that others may be better in other areas. Getting the USP right should provide a contractor with:

- a convenient rallying-call for company staff
- a basis to become 'top-of-mind' when prospective clients think about contracting services.

An effective marketing programme will need to comprise the following features.

- Prepare comprehensive listings, which are properly cross-referenced, of prospective clients in all chosen segments of the market. These listings should be supported with details on spending history and, whenever possible, future spending plans.

- Prepare listings of key decision influencers, particularly the consultants and public sector bodies. Again these listings should be properly cross-referenced and categorized by the type of work that would be bid.

- Maintain accurate and up to date statements of experience and qualifications.

- Install a worldwide capability to identify all relevant invitations to bid for work.

- Undertake a conscious programme to meet key prospective clients and decision influencers on a regular basis. The objectives of these meetings should be to:

  - get better acquainted
  - learn more about future spending plans
  - update people about latest experiences
  - develop relationships which will encourage prospective clients to discuss problems or factors which inhibit their plans, in the hope that an innovative solution can be found.

- Support these field activities with programmes which reinforce the marketing message of the business. Typically this will involve:

  - production of a brochure

- advertising
- entertaining.

These activities can be very expensive and the entertaining side needs careful management and sound judgment. Good documents and brochures are essential. All other activities should be determined on the basis of justified total cost (i.e. set the budget first, and design the programme second). Advertising is the most expensive way to make an impact, but it is a critical part of achieving a 'top-of-mind' position whereby key people would automatically recall your company name when asked about contractors.

The contracting business is almost legendary in its reputation for entertaining. However, entertaining is an area which needs careful thought and judgment and it is better to err on the side of doing too little rather than too much. In many ways it is a more powerful weapon to maintain existing relationships than it is to forge new ones.

Essential staff work

As with the engineering consultants, it is vital to undertake segmentation and business analysis exercises. Without this work marketing will be misdirected no matter how polished and professional it may appear.

Contracting as an activity may also be about to enter a new era in operational terms. There is greater pressure on margins at a time when some skills are experiencing shortages in supply. This is a classic recipe for investment in new equipment, automation or management control systems. The industry has been remarkably resistant to automation in the past, but the prospects for intelligent machines with programmable controls are starting to look encouraging. The successful contractors of the future will have put some investment into R&D, in ways not previously contemplated by the industry, to maintain a competitive edge on bidding for work.

The prospects for Venturing

Perhaps the biggest impact on contractors will come in the future from the need to:

- develop work with greater certainties
- provide longer term streams of income from some operations to offset the unpredictable nature of bidding successfully

- respond to the growing trend in the privatization of public sector operations.

One way of responding to these challenges is to develop the concept of Venturing in an organization. A venture constitutes a project where the contractor is willing to make an equity investment in the concept for design, financing and implementation and not merely look for a position to bid for work.

- The attractions of this activity are that:

- there is greater certainty in securing work once suitable venture projects have been identified and marketed;
- the activity provides a longer term stream of income from the equity investment and this improves the quality of earnings for a contractor by reducing the selling effort in subsequent years to maintain a given level of profits.

These benefits, however, are not easily won. Venturing is often a very long lead-time activity with considerable upfront investments needed in market development and engineering design. However, with the deregulation of public utilities this approach offers contractors an ongoing stake in infrastructure related projects and the prospect of being in more control of future civil engineering work awards.

15 The public service authority

Role

The primary mission of public service authorities is to provide and operate a service to the public in a specific area, such as water, gas, or power. Their effectiveness is often judged on the way in which the public perceives the product rather than the means by which the product is produced or delivered.

In the UK they are often organizations that both purchase services from consulting engineers and contractors, and carry out work on their own account with their own resources.

They are often required to orchestrate the whole process of:

- identifying public demand in their sector;
- providing sufficient capacity to meet public demands;
- organizing consulting engineers and contractors to provide facilities and services;
- managing facilities professionally to provide public services.

Evolution of the profession

- The driving force behind the creation of public service authorities, historically, was the desire:
  - not to duplicate expensive infrastructure in meeting public needs;
  - to provide authorities which were accountable to the general public for the provision of key services.

Many public service authorities relate to the provision of some type of utility service or other, such as water, gas, electricity, motorways and roads, etc.

As the public sector organizations were being set up in the UK, there was a general move towards greater economies of scale in operation and very often greatly increased public demand for services. The public service authorities found that they were in a perfect position to solve the funding problems that the private sector often struggled with, with infrastructure, by using government or equivalent guarantees judiciously.

This led to a much vaunted approach to public services in the UK and the growth of a class of professionals to run the various authorities. These professionals, however, are not typically responsive to normal commercial pressures. The two important operational objectives in the public sector are:

- to minimize the number of occasions when there is a shortage of supply;
- to develop operations in a fully accountable way to the elected members of the public authorities and to the public in general.

This approach often leads to solutions that are conceptually very different from those developed in the private sector of the economy.

Future trends

The public sector is likely to experience more change in the near future in relation to civil engineering than will either consulting engineers or contractors.

Constraints on public spending have become a worldwide phenomenon, and the UK has become one of the leaders in exploring the privatization route for reducing tax burdens and public spending. These trends are having a major impact on all public sector authorities, particularly in questioning fundamental business and commercial objectives. On present projections the 1985–95 decade will see enormous structural changes in the whole public sector field.

Many public servants will become operations managers in new commercial enterprises. They will be competing with some of those contractors that plan to follow the venturing approach to secure a position in the fields of public infrastructure and utilities. The key attributes of the public servant in these circumstances will be operational experience and a commitment to service excellence. The main areas of vulnerability are likely to be cost competitiveness and a lack of sensitivity for customer service issues.

The stage is being set for many interesting developments in this whole area of public services.

The key results areas

- The public sector authorities have four major groups to market to:

  - the public, for sales of product;
  - politicians, for an acceptance of their role in the economic activities of the UK;
  - prospective customers, for the purchase of professional and operational services based on in-house expertise and which could be made available to others on a commercial basis;
  - prospective partners from the private sector, to develop a controlled response to the pressures of privatization.

Marketing to prospective customers and prospective partners is very much part of the scenario that will face civil engineers in public service authorities in the future.

Key marketing themes

The civil engineer in the public sector may find that rather than having to market a product or service, they are more often the target of marketing efforts from consulting engineers and contractors. Often the public servant is a major purchaser. In this context the public servant should be mindful of the marketing approaches of both the consulting engineer and the contractor in order to secure the results he seeks in his own capacity.

However, there is an extra marketing role to be fulfilled by public servants in relation to the public, politicians, prospective clients and prospective partners. There is an important role in image forming and in the dissemination of information to the industry. Marketing in this context has a very high public relations content.

An effective marketing programme will need to comprise the following features.

- Publish well prepared brochures and materials to help people be more knowledgeable and more comfortable with operations.
- Prepare comprehensive listings of all clients with referencing to the services they take and the performance achieved in meeting service requirements.

- Undertake a structured programme to meet all key clients on a regular basis to discuss:

  - future plans for their operations;
  - ways in which more appropriate or more cost effective services should be supplied in the future;
  - the experience of their international competitors in securing the sorts of services you have been offering;

- Identify those areas of activity where the concept of contract services can be developed for third party operations or for similar public sector authorities around the world. Attention should be focused on those areas where:

  - a clear leadership position is held in terms of experience or technology;
  - extra activity will help to support the cost of specialist in-house resources not available from the private sector.

Essential staff work

With absence very often of commercial management objectives, the public sector authorities should develop programmes to:

- ensure that services maintain a high level of availability at all times;

- ensure that the unit cost of the services provided is as attractive as possible through:

  - tight control of operational costs
  - minimization of capital expenditure consistent with meeting operational standards.

16 Conclusions

Contrasting characteristics

There is no single approach that can be advanced for civil engineers to undertake marketing. Not only will there be important differences in detail from one situation to another, but there are some fundamental differences in the roles that engineers play in their respective organizations.

- Engineers in academia are charged with advancing the state of the art and communicating progress to the rest of the profession.
- Consulting engineers are concerned about developing appropriate concepts and securing the best performance possible from engineering aspects.
- Contractors are critical to taking on the burden of risk and will seek leverage through productivity, and the efficient use of appropriate equipment.
- Public service authorities have the onerous burden of public accountability and are committed to providing an optimum level of service within any imposed budgeting constraints on their activities.

The professional civil engineer can therefore find that he may be required to fulfil a wide variety of roles, according to the organization he represents, and these roles will bring distinctive marketing requirements.

These contrasting characteristics are summarized in Fig. 29.

Pattern for career development

The biggest barrier to effective, professionally-performed marketing in civil engineering is the attitude of the engineer. Most engineers

Organization	Primary roles	Key differentiating features
Academia	Researcher and trainer	Best materials Best methods
Consulting engineers	Designer and auditor	Best performance Best concepts
Contractors	Doer and risk taker	Best equipment Best productivity
Public service authorities	Buyer and operator	Best service Full public accountability

Fig. 29. Contrasting characteristics

wish they did not have to market their business at all, and that there was some mechanism that could be set-up which would remove the need for it.

This problem has to be tackled at its roots and it is worth starting with a review of career development in the profession.

The engineer, in pursuing a career to the very top, should go through four stages of evolution:

- *Technical specialist.* In this phase the engineer develops his technical competence and his base of experience. He will then be faced with making his next step.

- *Man manager.* All civil engineers are exposed to the best, toughest and most rewarding man management opportunities in the business world when they embark on a career as a project manager. It is often said that those who revel in this become contractors, those who miss the technical aspects become consultants, and those with a highly developed sense of hands-on development join public service authorities.

  Unfortunately, most careers are not thought through to beyond this phase.

- *Businessman.* In this phase the engineer is not only involved in concluding business arrangements, he is also involved in generating new revenues. This phase of development tends to come earlier to consultants than to contractors or engineers in public authorities.

This is where marketing starts to play an important part and the profession typically is not well prepared to make this step emotionally. Consequently, success as a businessman is very hit-or-miss.

- *Entrepreneur.* This phase is often forgotten completely, but the ultimate mission of the professional businessman is to make things happen by risking his resources and backing his judgment. All engineers should aspire to this position.

By adopting a positive long-term attitude towards career development, the civil engineer will see the relevance and excitement of marketing and will justify a structured approach to doing it well.